Plantations, Pineland Villages, Pinopolis and Its People

Proceeds from the sale of this book have been assigned to the Berkeley County Historical Society. A portion of those funds will be used to restore the old Pinopolis Post Office.

THE
DONNING COMPANY
PUBLISHERS

LAKE MOULTRIE
Old Ditch Line Location
Old Ditch Line Location
Existing Ditch Line
No Longer Existing Homes and Structures
A Wesley United Methodist Church
B H.W. Harvey House and Store
C Peterson House
D Pinopolis Crossing Station House
E Cotton Gin
F Methodist Parsonage
G J. Calhoun Cain (same site as 11)
H Cuthbert Hare Grits Mill
I Cuthbert Hare Store
J Thomas W. Williams House
K Fredrick A. Porcher House
L Pinopolis Public School
M Hawkins Jenkins House
N Tennis Court Lot
O Ophir Summer House
P Ophir Tennis Court
Q Old Trinity Episcopal Church
R Samuel P. Stoney House
S Woodboo Summer House
T Rene Ravenel House (same as 28)
U Whitehall Summer House
V Sarazins Summer House
W William Henry Cain House
X John P. Harvey House
Y Harlock Huxford Harvey House
Z Nazareth School
AA Nazareth Church
BB Cedar Spring Summer House
CC L.H. Harvey / Murray House
No Longer Existing Homes of Black Families
1 McKinley and Eliza Smalls & Mary Sheppard Smalls (Dolly) Brown
2 Second Home of Eliza and Mary Sheppard Smalls
3 William and Rosalee Perry House
4 Catherine and Ed Beatty House
5 Missouri Middleton and her daughter "Stay"
6 Mary Jane Smalls "Cook"
7 Franklin and Lucille Brown
8 Eddie and Ellen Burch
CHICORA DR
LAST LN
ALMOND ST
WAMPEE DR
VALMAR DR
RECREATION DR
SHORESIDE CT
NEW ENGLAND DR
COVE COTTAGE LN
BROUGHTON HALL RD
QUAIL DR
OLD VILLIAGE DR
FOX SQUIRREL LN
DODGE LN
LAKENGLEN LN
MORNING DOVE LN
TRANQUILITY LN
LUCAS ST
BAY LN
WALSH DR
0 1,000 Feet
Courtesy of Berkeley County Government
Geographic Information System Department

Plantations, Pineland Villages, Pinopolis and Its People

Norman Sinkler Walsh, MD,
Author and Photographer

Research by Cecy Guerry

Second printing 2007

The Donning Company Publishers
184 Business Park Drive, Suite 206
Virginia Beach, VA 23462

Steve Mull, General Manager
Barbara Buchanan, Office Manager
Pamela Koch, Editor
Lynn Parrott, Graphic Designer
Chad Casey, Reprint Graphic Designer
Mellanie Denny, Imaging Artist
Derek Eley, Reprint Imaging Artist
Scott Rule, Director of Marketing
Tonya Hannink, Marketing Coordinator

B. L. Walton Jr., Project Director

Library of Congress Cataloging-in-Publication Data

Walsh, Norman S. (Norman Sinkler), 1935-
Plantations, pineland villages, Pinopolis and its people / Norman S. Walsh, author and photographer ; research by Cecy Guerry.
p. cm.
Includes bibliographical references and index.
ISBN-13: 978-1-57864-385-1 (hard cover : alk. paper)
ISBN-10: 1-57864-385-6 (hard cover : alk. paper)
1. Historic buildings--South Carolina--Pinopolis. 2. Historic buildings--South Carolina--Pinopolis--Pictorial works. 3. Dwellings--South Carolina--Pinopolis--History. 4. Plantations--South Carolina--Pinopolis--History. 5. Pinopolis (S.C.)--Social life and customs. 6. Pinopolis (S.C.)--Biography. 7. Pinopolis (S.C.)--Buildings, structures, etc. 8. Berkeley County (S.C.)--History, Local. 9. Historic buildings--South Carolina--Berkeley County. 10. Plantations--South Carolina--Berkeley County--History. I. Guerry, Cecile Ann. II. Title.
F279.P56W35 2006
975.7'93--dc22

2006025752

Printed in the United States of America at Walsworth Publishing Company

Contents

Dedication

Elias Ball Bull, 1930–1994

Elias Bull. *Courtesy of South Carolina Historical Society.*

Elias Ball Bull and his parents, Julia and Kinloch Bull, moved to Pinopolis in 1948. Elias attended the University of New Mexico and the University of South Carolina, where he received a bachelor of arts degree with majors in archeology and political science. He had various jobs in Berkeley County, including social caseworker, a part-time job in the Pinopolis Post Office, and clerical work.

Robert Stevens, Jimmy Skinner, and I, as teenaged boys in the 1950s, were intrigued with Elias because he was about five years older than we were and told interesting stories about archeology. Due to Elias's influence, Robert and Jimmy were stimulated to collect artifacts that were exposed along the shores of Lake Moultrie. In Robert's boat, they visited remote islands and beaches. Elias was particular in teaching Robert and Jimmy to record and document their findings. Robert remembers that Elias became an authority on English pipes that were traded locally to the Indians.

Elias was a life member of the South Carolina Archeological Society and a member of the South Carolina Poetry Society and the South Carolina Historical Society. He will be remembered as a prolific and thorough researcher and organizer of Berkeley County history. In his capacity as Historic Preservation Planner for the Berkeley-Charleston-Dorchester Council of Governments, Elias completed several inventories of historic preservation in Berkeley County and took numerous photographs of historic buildings. These photographs and other documents are currently housed in the Charleston County Library and the South Carolina Historical Society.

A memorial article about Elias Bull in *Carologue*, a publication of the South Carolina Historical Society, written by Harlan Greene, archivist for the Avery Research Center for African American History and Culture in Charleston, described Elias as: *by turns, warm and ironic, contrary, and genteel. He was modern in outlook, but proud heir to many "old fashioned" legacies, such as politeness, gallantry, and a sense of duty . . . His utter devotion, his unceasing research, his curiosity and honesty will be his legacy. Unlike many historians, he was not so much desirous of a publication as he was with knowledge itself; it was not the pomp of a completed destination he was after, but the learning of the journey.*

This book would have been incomplete without the work and records left by Elias Ball Bull.

Margaret Edwards Stevens, 1900–1982

Margaret Edwards Stevens. *Courtesy of Robert J. Stevens Sr.*

Margaret Edwards Stevens was the daughter of Rev. Joseph E. Edwards and Eleanor Willis Brown Edwards, who owned Chelsea Plantation summer house in Pinopolis. When Margaret married Goodloe Stevens in 1924, her parents gave them the Chelsea summer house lodge as a wedding gift. Margaret later inherited the Chelsea summer house.

Highly educated, Margaret completed the Spartan Academy in Spartanburg and Furman University. She attended Mason's Business College in Augusta, Georgia, Murray Vocational School in Charleston, and the School of Ceramics at the University of South Carolina. She taught school in Walhalla, South Carolina, and in Berkeley County. She was a staff writer for the *Berkeley Democrat* and the Charleston *News and Courier* and did freelance writing. She was also a genealogist, pianist, and, at one time, first violinist for the Charleston Symphony Orchestra.

Margaret was the author of numerous historical essays and had an extensive collection of documents regarding Pinopolis history. It is likely that Margaret had intended to write a history of Pinopolis. Her research was the foundation for "Pinopolis: History of a Pineland Village" by Anne S. Ball and for this book.

Marie Harvey, 1903–1988

Marie Harvey. *Courtesy of Gwen Lewis.*

Marie Harvey was born in Berkeley County, the daughter of Lewis Henry Harvey and Minnie Elizabeth Beatson Harvey. The family moved from Longridge in Berkeley County to Pinopolis in 1912. Her father built the existing home known as the Lewis H. Harvey house. Marie graduated from the Pinopolis School and Winthrop College. She later earned a master of education degree from the University of South Carolina. Marie Harvey taught third grade at Berkeley Elementary School from September 1924 until June 1970. She then assumed the position of Supervisor of Instruction at Whitesville School, which she held until her retirement in July of 1973.

She was my teacher in 1944. My parents and teachers were frustrated about my poor handwriting, and "Miss Harvey" was hired to tutor me in writing once a week after school. I rode my bicycle to her home, the Lewis H. Harvey house, where she had a table prepared in the front sitting room. On this table were sample books of script, which, under her direction, I attempted to copy. She finally gave up and said, "You might as well become a doctor like your father."

Miss Harvey was a charter member of both the Moncks Corner-Pinopolis Book Club and the Town and Country Garden Club. She was a member of the Berkeley County Library Board, serving as the secretary/treasurer from its organization in 1934 until 1964, when she became chairman of the board. During her term as chairman, two branch libraries were organized in Hanahan and Goose Creek. A new library building was constructed in Moncks Corner, opening in 1970. She retired from the library board in 1981 and wrote a book titled *History of the Berkeley County Library 1934–1981.*

In 1983, Miss Harvey assisted Anne S. Ball in writing "Pinopolis: History of a Pineland Village." It is likely that she compiled and organized the records of Margaret Stevens and participated in new research. Miss Harvey retained some of these records, which, at her death, were left with her niece, Gwen Lewis of Pinopolis.

Anne Sinkler Fishburne Ball, 1913–1983

Anne S. Ball. *Courtesy of Anne Ball Helms.*

Anne Sinkler Fishburne, daughter of Anne Sinkler Fishburne and William Kershaw Fishburne, MD, was called "Peach" by her father. From then on, Peach became her name. She went to the Pinopolis School, Ashley Hall, and the University of Pennsylvania. She married William Moultrie Ball in 1935. During World War II while Moultrie was in the army in the South Pacific, Peach lived in Pinopolis with her parents. During the gas shortage, she rode a bicycle from Pinopolis to her work in Moncks Corner and back each day.

Peach was active in the Berkeley County Historical Society, the Berkeley County Library, the Town and Country Garden Club, and the Moncks Corner-Pinopolis Book Club; she also worked with 4-H clubs (an organization providing educational opportunities for rural youths) in Berkeley County.

During the time that her husband, Moultrie, was administrator of the Berkeley County Hospital, Peach worked there in various capacities.

A lifelong member of Trinity Episcopal Church, she sang in the choir and volunteered in all of the women's organizations. Peach and Moultrie had one daughter, Anne Ball Helms, who lives with her husband in Charleston. Peach was my cousin, friend, and godmother of our daughter, Louisa.

Anne S. Ball's printed but unpublished book, "Pinopolis: History of a Pineland Village," was one of the stimuli for this book and a valuable source of information.

Preface

To drive the old roads, past houses where people I loved used to live, is almost unbearable. What my eye sees is not what my heart seeks. Which of course, is everything put back together the way it used to be. . .only better.

Nikki Hardin, *Skirt*, September 2005

When I read those words in *Skirt* magazine, I felt a twinge of sadness. I suspect that it is common for those who go back to the places where they have grown up to get that feeling. When I was growing up in Pinopolis in the 1940s and 1950s, the village was in many ways like it had been for the previous one hundred years. It was a small close-knit community. Everyone knew each other and cared for each other because, for the most part, we were all part of each other's extended family. At the time, I didn't appreciate being caught for some misdeed and disciplined by neighbors. Now, I have the utmost gratitude for their guidance and genuine caring.

There was a network of kinship among the early families: Porcher, Cain, Ravenel, Dwight, Lucas, Hare, and Harvey. Visiting back and forth among neighbors was a common occurrence, children and adults shared mutual respect, and life was slow paced.

The village was shaded by mature longleaf pines, and roads were no more than sandy paths winding from house to house through the woods. The rustic, old homes were plain and simple, hot in summer and cold in winter. The weathered pine or cypress siding of the houses was commonly unpainted. Other than private homes, the village contained only one small store run by the Hare family, three churches for whites, and one church for blacks.

Today, the wilderness spaces, bare wood homes, and giant pines are gone. No one would want to live in the homes of Pinopolis as they were in the 1940s and 1950s, and I wouldn't either.

Although still a village, Pinopolis is no longer a community of extended families. There are no Lucas, Porcher, or Ravenel families remaining in Pinopolis; however, descendants of the Hare, Harvey, Guyton, Dwight, and Cain families still live there. Pinopolis maintains a village atmosphere as it is a small enough place that neighbors know each other and children safely play outside. Since the creation of Lake Moultrie, Pinopolis has one main road through the peninsula—only one way in and out. The village has never been incorporated.

I was born in 1935 in Charleston and lived my first year on Broughton Road in Moncks Corner with my parents, Dr. Joseph Norman Walsh and Cleremonde Gaillard Sinkler. I called Pinopolis home for the next twenty-two years until 1958 when I married Louisa Marcy Stoney of Charleston. My first home in Pinopolis was the house known as the Episcopal Rectory. When I was six, my parents acquired what was known as the Causey property. In 1942, my father built the house presently known as the Walsh/Russell house. I remember standing with my father in an abandoned field in our front yard watching Lake Moultrie's waters rise before me. Once completed, this rising defined the boundary of Pinopolis as it is today.

I have lived in other places for almost fifty years, but Pinopolis still has the feeling of home. I attempt in this book to preserve some of the past with words and photographic images.

The actual stimulus for this book was technology, not nostalgia.

In the fall of 2003, I became interested in digital photography. While visiting in Pinopolis, I did feel nostalgic about the old homes, buildings, and people of the past. I saw the places where fifteen old structures, present during my childhood, no longer existed. I had a desire to make a photographic record of existing historic structures.

Through this book, I intend to: (a) present an overview of pineland villages of Berkeley County; (b) portray life in a pineland village, specifically Pinopolis; (c) give an updated history of Pinopolis homes and structures; (d) display photographs, old and new, of Pinopolis village homes and their corresponding plantation homes that no longer exist; (e) showcase as many images as possible of structures no longer existing. Copies of photographs and research materials will be donated to the South Carolina Historical Society.

This book links plantations long lost to the waters of Lake Moultrie with their associated summer homes in Pinopolis, which existed in its early years solely as a summer retreat from the malarial fevers of the plantation environment. Described here are thirty-four historic structures in existence in 2005 and twenty-six that no longer exist.

There have been three significant historical writings about Pinopolis. The first was in the memoirs of Frederick Adolphus Porcher, one of the first two settlers of Pinopolis. Richard D. Porcher, PhD, wrote an excellent history of Pinopolis for a high school senior paper in 1957 (unpublished). In 1983, Anne Sinkler Fishburne Ball privately printed her book "Pinopolis: History of a Pineland Village." She relied heavily on information in Porcher's manuscript and went into detail regarding individual homes. She described twenty-eight buildings in existence then and eight known structures that no longer existed. Anne S. Ball's book contains many interesting stories about village life that are well worth reading.

By contributing stories and images in this book, I endeavor to preserve the memory of Pinopolis—its people, their way of life, and their homes. Proceeds from sales of this book will be donated to the Berkeley County Historical Society.

Norman S. Walsh
Summerville, S.C.
May 2006

Acknowledgments

I had so much cooperation and assistance in gathering information for this book that there is not enough space to share how each person helped. The homeowners were gracious in giving me unrestrained access to their homes. Their names are listed at the end of each home's history.

Many others contributed interviews, photographs, old family documents, and useful information: Robert Austin; Beth Bilderback at the Caroliniana Library; James Bunch and Francis Rackley Bunch; Mary Brown; Maude Smalls; Steve Buckler; Nicholas Butler at the Charleston County Library; College of Charleston Library; Joe Cross; Edgar Davis; Katherine Davis deLoach; Henry Ravenel Dwight Jr.; Herb Frazier; Sandy Gibson; Harlan Greene; Dewey Ford, Dozier Helmly, and Bobby Helmly at the Home Telephone Company; Cynthia Hamilton; A. Davis Hare Jr.; Dan Hare; Harold H. Harvey Jr.; Anne Ball Helms; Camilla Bishop Henry; Samuel Heyward; Van High; James M. Huff of Santee Cooper; Colleen Lacey; E. Louise; Dorothy Massalon; Joseph H. McGee; Marianne Davis Monroe; Ophelia Morris; Laverne Mazyck; Larry Motes; Mildred Myers; Mildred Owens; Emily Palmer; Catherine Rackley Parler; Rosalee Perry; Beatrice Dennis Plyler; Heyward Ravenel Porcher; Margaret Deas Porcher; Percival Ravenel Porcher IV; Richard Dwight Porcher Jr. PhD; Martha Mitchell Porcher Ravenel; Sandy Sanders; Virginia Seay; Ellison and Celeste Singletary; William and Justine Smalls; Louann Brown Smith; Mike Coker, Karen Stokes, and Eric Emerson at the South Carolina Historical Society; Sarah Spruill; Robert Stevens; Albertha Varner; Julia Parker Young; John Waddill; Waring Medical Library; and Margaret Waters.

Bernie Walton, project director for the Donning Company, has been intimately involved in this project from the beginning. I am grateful for his generosity of time, advice, encouragement, and good humor. He has become a good friend as well. I have special appreciation for my editor, Pamela Koch, whose availability, knowledge, encouragement, and gentleness eased this novice author through the process. Graphic designer Lynn Parrott has amazing creativity as well as a gift for arranging images and text in a most appropriate and appealing manner.

The Berkeley County Historical Society is the recipient of the profit from this book. The Board of Directors of the society, and especially, its president, Anne Propst, has been patient with me for my tunnel vision about this project.

Mike Holcombe of the Farmers and Merchants Bank relieved our anxiety about going into debt by underwriting the project. I am grateful to the members of the "Book Committee" for their individual contributions and for being available whenever I asked for help: Elaine Swicord, Samuel Pledger Parker Jr., Gwen Lewis, Dora Wiley, and Cecy Guerry.

I have great gratitude and respect for Cecy Guerry for her research and dedication to this project. To her goes the credit for the historical detail and all information that supplants that in Anne S. Ball's book, "Pinopolis: History of a Pineland Village." In addition, she has edited every word of the finished product.

My wife, Marcy Stoney Walsh, has supported me every step of the way. She has played an active role as an editor, has created a useful index, and has helped me in my ignorance of Microsoft Word. She was unfamiliar with the history of St. John's Berkeley and Pinopolis. By constantly asking me to explain that history to her, I came to realize the need to do the same for you, the reader.

I apologize to anyone I left out. Let me know, and you will be in the next edition.

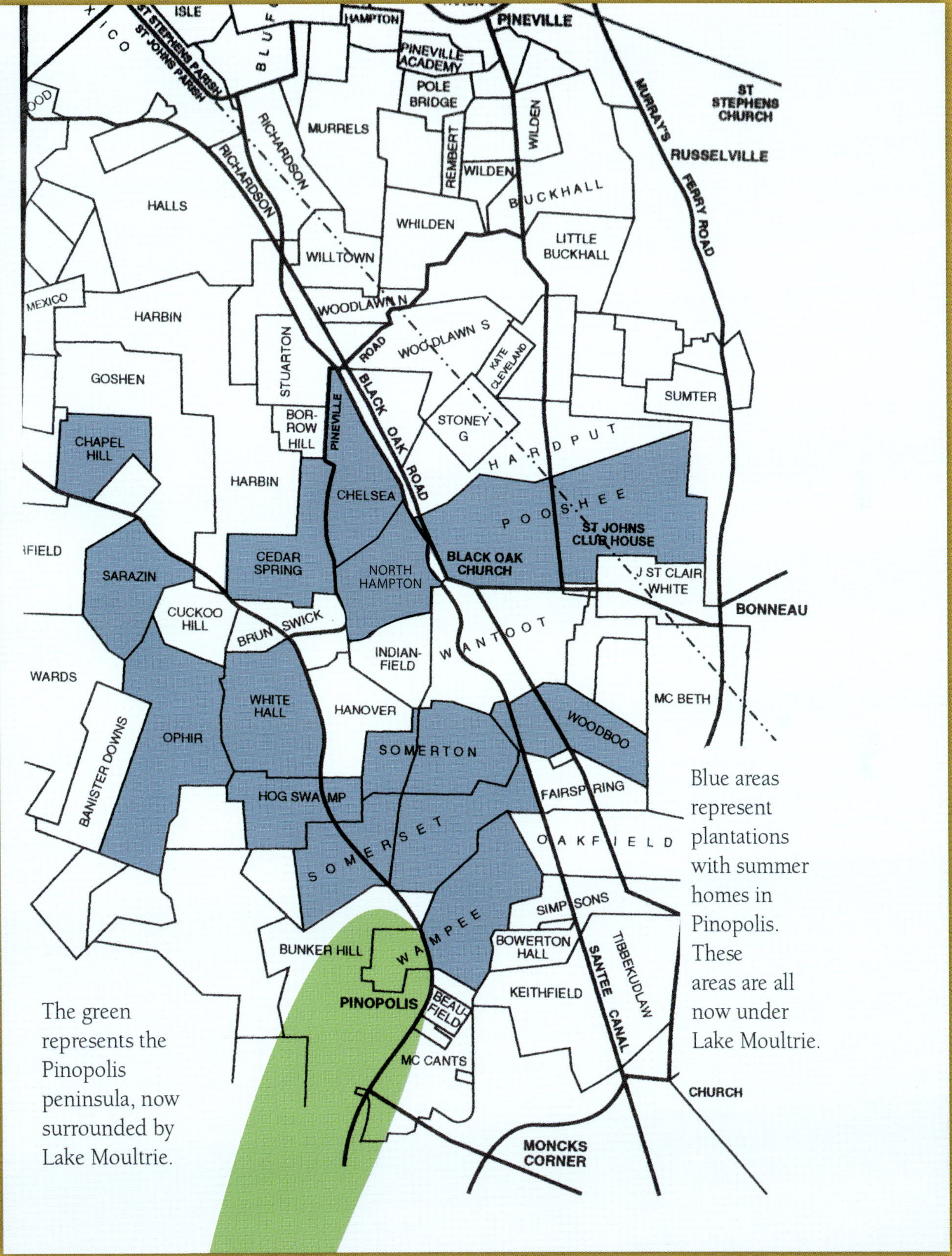

The green represents the Pinopolis peninsula, now surrounded by Lake Moultrie.

Blue areas represent plantations with summer homes in Pinopolis. These areas are all now under Lake Moultrie.

Plantations of St. John's Berkeley that had a Pinopolis summer house.
Adapted from a map by John Palmer Gaillard.

Introduction

Perspective

A story placed in historical perspective is easier to understand. The Vikings explored the northeastern American continent about 1000 AD. Spain had a fort on Parris Island, South Carolina, in 1566. The first permanent settlement in America was in 1607 at Jamestown, Virginia. In 1663, King Charles II of England gave a land grant, which included what is now South Carolina, to eight English entrepreneurs who were called the Lords Proprietors. The legal charter gave the Proprietors title and authority over the territory of "Carolina," which was not divided into "North" and "South" until 1729.

Charleston was settled in 1670. In 1682, the Proprietors created the counties of Craven, Berkeley, and Colleton inland from Charleston. Craven County was settled in 1689 along the Santee River near Jamestown by French Huguenots, who were Protestants seeking religious freedom from Catholic persecution (Savage 1968, 101).

Berkeley County was named for two brothers, Sir William (1606–1677) and Sir John (1607–1678) Berkeley. In 1798, Berkeley was incorporated into Charleston County and remained there until 1893. The move to re-create Berkeley County, begun in 1882, was an intentional act by Charleston Democrats to keep political control in the Charleston area. This change put most of the Republicans, the majority of whom were Negroes, outside of the city and county of Charleston. After that, the Charleston newspapers referred to Berkeley as the "black county." At the time of the separation from Charleston County, Mount Pleasant was the seat of Berkeley County. In 1895, the seat was changed to Moncks Corner (Orvin 1950), which had been settled between 1735 and 1747.

South Carolina was also divided into parishes. This causes confusion for the casual reader of South Carolina history. The counties can be pictured as geographical entities and the parishes as political and religious (Church of England) structures within the county and state. The Church Act of 1706 established the parishes of St. John's Berkeley; St. Stephen's; St. James Goose Creek; St. Thomas; St. Dennis; and St. James Santee. These were not geographic boundaries; they were Church of England parish divisions, which became social segments for the planters and election districts for the Commons House of Assembly.

The approximate area of St. John's Berkeley is described by Frederick A. Porcher, a prominent planter and writer, as follows (1943, 67): *St. John's parish began on the South at the point where the Back river flows into the Cooper river and extends in a direction a little west of north until it reaches the Santee river near the Santee Canal, and borders the river westward until it reaches Nelson's Ferry* [near Eutaw Plantation]. *It is thus about twelve miles in width and about fifty in length. at its lower end it is hemmed in between the waters of Cooper & Back rivers, but its eastern boundary is near Huger's Bridge. Thus this parish included a large amount of the best lands in the low country. It contained all the Rice lands of the western Branch of the Cooper River, and all on the western side of the Eastern Branch; all the cotton lands that border on the streams that flow into these Rivers, the most important of which is Begin creek and its numerous tributaries; and all the rich limestone soil that borders on the north of the Santee Swamp, and which proved so eminently adapted to the culture of Cotton.*

Harriette K. Leiding further describes the parish divisions (1921, 132): *It is a strange thing to note that there are three very arbitrary divisions of St. John's, not easy for an outsider to understand . . . No more puzzling occupation can be devised than to correctly place the different families of the same name in their correct places. Suffice it to say, that the settlement in Upper St. John's was called Eutawville, where several houses are still found . . . Middle St. John's settlement is called Pinopolis . . . in Lower St. John's were the summer settlements of The Barrows and Cordesville.*

Eutawville in Upper St. John's was transferred to Orangeburg County in February 1910. *Orangeburg County also acquired a rich section of Berkeley territory lying west of an irregular line running slightly northeast and north from the confluence of Dean and Four Hole Swamps to the point where Rock Creek empties into the Santee River. In this area were the towns of Holly Hill and Eutawville, and the famous Eutaw Springs* (Orvin 1973).

This book focuses on Middle St. John's parish, which had active plantations and homes in the early 1700s. The lowland, with its fields and plantation homes, was the most fertile but is now under Lake Moultrie, leaving only the remnants of the pineland villages to remind us of lost places and long-ago times.

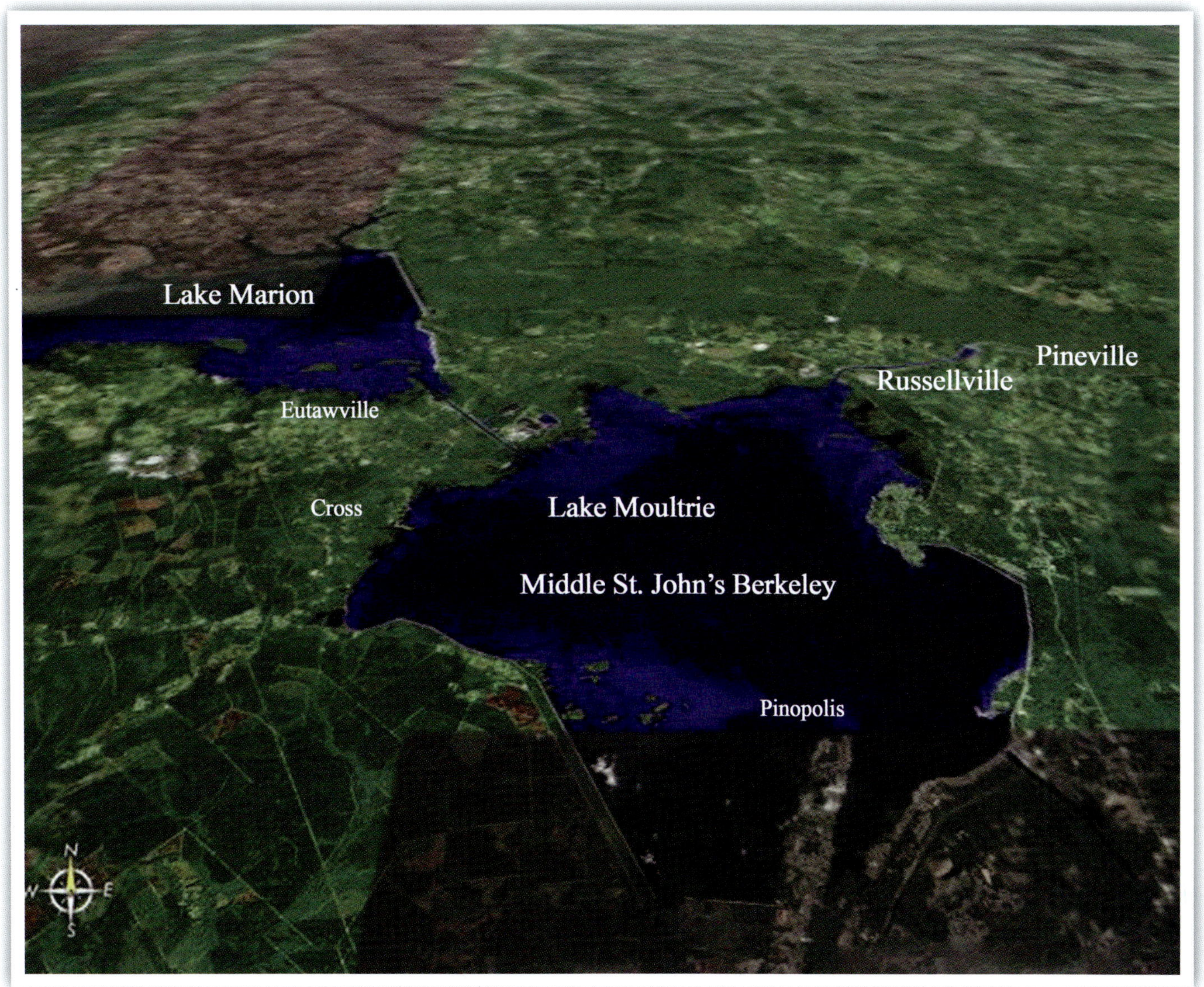

Aerial view shows that Middle St. John's Berkeley is now under the waters of Lake Moultrie. Pinopolis still exists because it is a sandy peninsula of high ground. *Courtesy of Google Earth.*

chapter I

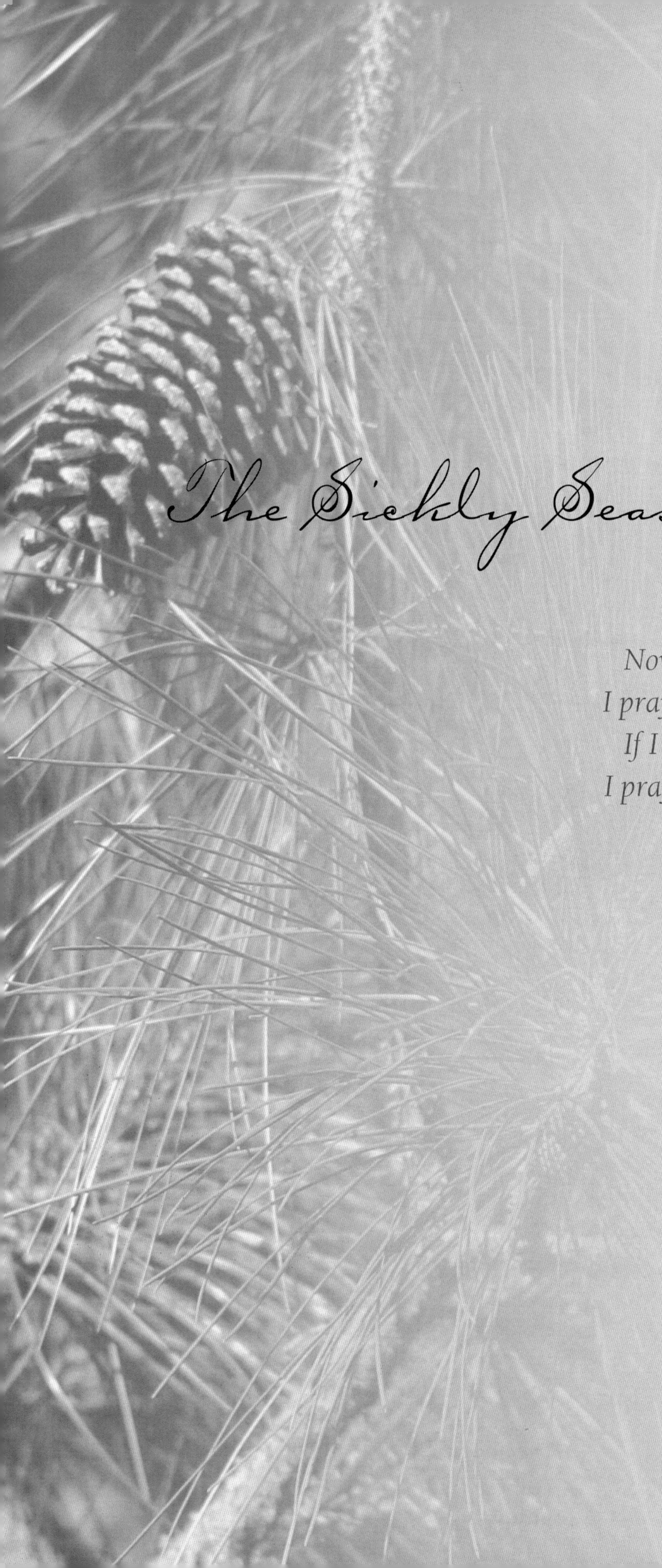

The Sickly Season, Malaria

Now I lay me down to sleep;
I pray the Lord my soul to keep.
If I should die before I wake,
I pray the Lord my soul to take.

FOR ALMOST ONE HUNDRED YEARS, PLANTATION LIFE flourished in Carolina without malaria being a significant issue. In the late 1700s, something changed. The death rate from malaria increased noticeably. There came a time when the eighteenth century prayer opposite could have been said in every household of St. John's. The threat of malaria changed the planters' way of life. If this disease had not existed, there would have been no need for plantation families to begin building homes in the pineland and living there from May to November. By providing a safe haven from disease, the summer migration to the pineland allowed the plantation system in St. John's to continue from 1794 until the Civil War in 1861.

The plantation was a delightful place to live. Its residents would not have subjected themselves to the semiannual move, to and from plantation to pineland village, unless their lives depended upon it. It is still controversial as to whether malaria was endemic or later imported to the Americas. However, there is no controversy concerning the vital role that pineland villages played in escaping malaria.

Frederick A. Porcher described the discovery of the pineland as a healthy place in his extensive memoirs (1887, 53–54): *After the year 1790, when freshets in the river* [floods along the

Santee River] *became more frequent, the climate became more sickly. The residents along the swamp suffered severely from agues and fever . . . Capt. James Sinkler, who was a sagacious observer, was led from his observations to believe that a pineland residence, even but a short distance from the swamp, would secure its occupants from fever. Acting on this notion, he built a house for himself in the pineland, and in June 1793, retreated to it with a family, blacks and whites included, of more than twenty persons. In November, he returned to his plantation, having passed the summer in the enjoyment of uninterrupted health. This experiment was immediately imitated* [in the settling of Pineville, one of the earliest pineland villages] . . . *a suffering people quickly became contented and happy.*

The planters chose to farm the lowland because of its fertile soil. Unfortunately, this is where mosquitoes breed. The sandy soil where longleaf yellow pines thrived was higher ground that quickly drained water. It was not until one hundred years after the settlement of Pineville, in 1794, that the mosquito was known as the carrier of malaria; but the discomfort from this pesky insect was often remarked upon in letters and journals. As early as 1725, wealthy settlers were using protective mosquito netting made of catgut.

William Cain (1908–1993), living at Somerset Plantation and Pinopolis in the 1920s, remembers: *It* [the plantation house] *was located quite near the swamp and mosquitoes would have rendered it absolutely uninhabitable during a large part of the warm weather.*

Extensive reading about the "sickly season" and "summer migrations" arouses curiosity about malaria and its origin. The lifespan of plantation dwellers, especially the children, was cut short by many diseases; however, the most deadly was malaria. Malaria was also known as ague, miasma, country fever, swamp fever, intermittent fever, and, near the coast, myrtle fever.

There are many written accounts asserting that plantation life was healthy year-round before the American Revolution. In his book *Summer Migrations and Resorts of South Carolina Low-Country Planters*, L. F. Brewster (1947, 3) notes that planters had not always left their plantations during the "sickly season." One such planter, F. A. Porcher, said that his grandfather remained on his plantation throughout the year and was even visited every summer by his Charleston relatives (F. Porcher 1943, chap. 2). Samuel DuBose, another early planter of St. John's, stated clearly in "Reminiscences of St. Stephen's Parish," that in the beginning, the swampland was a healthy place to live: *About twenty years before the Revolutionary War, the belt of land bordering on the Santee River, through the whole extent of the parish of St. Stephen's, was the garden spot of South Carolina . . . A breadth of three or four miles of swamp as fertile as the slime of the Nile could have made it, was safe for cultivation; and its margins were thickly lined with the residences of as prosperous a people as ever enjoyed the blessing of God. Some there lived in the swamp, and even on the very bank of the river. The exceeding fertility of the soil rendered labor scarcely necessary to make it a wilderness of vegetable luxuriance . . . before the eye was spread nature in all her majesty and beauty . . . It was my delight, when a boy, to hear told how at evening the family would sit in the humble porch, and enjoy the rich delights of a spring twilight, listening to the songs* [of birds] *. . . the buzz of the bees . . . the sad wail of the whippoorwill . . . and admiring the gorgeous colors of the trees, shrubs, and plants* (DuBose 1887).

DuBose relates that the settling of the inland areas of South Carolina, the cutting of trees, and the plowing of the soil reduced the evaporation of water by trees, decreased the absorption of water by the soil, and increased the volume of spring floods (freshets) on the Santee. He contends that plantations, because of the changed conditions, became less healthy after 1790.

David Ramsey, a South Carolina historian, planter, and physician, attributed the change to rice planting: *sluggish rivers, stagnant swamps, ponds, and marshes are common; and in or near to them putrefaction is generated. In all these places, and for two or three miles adjacent to them, the seeds of febrile diseases are plentifully sown and from them are disseminated* (Ramsey 1809).

Robert Mills, who was collecting information on the natural history of South Carolina, stated in 1826 that in summer and fall the swamps retained water, thick fogs occurred during the night, and those who were exposed to the fogs would get intermittent fever (Mills 1826).

On November 28, 1843, Dr. S. H. Dickson addressed the South Carolina State Agricultural Society and presented views held by his contemporaries: *The beautiful and fertile low country of our State is the seat of annual and endemic visitations of disease, which we are accustomed to attribute to Malaria. Whatever may be the difference of opinion elsewhere as to the source of origin of the aerial poison, the Medical Profession here is unanimous in regarding it as the result of vegetable decomposition in moist places at a high temperature . . . in the low country there are few plantations which admit of permanent residence; the whole region being pervaded by a pestilential infection, almost unfailing in the excitement of Fevers, Interment and Remittent, during the Summer and Autumn . . . the cultivation of rice . . . will always be found incompatible with the salubrity of the country engaged in it . . .* [and] *for obvious reasons, improvident clearing, imperfect cultivation, and neglected drainage of land, have been the curse of our middle and lower districts* (Dickson 1846).

In "Native Americans of North America" from Microsoft Encarta Online Encyclopedia, it is stated that malaria was introduced to the Americas in the 1500s. In *The Malaria Capers*, Robert Desowitz states that no references to malaria could be found in Mayan and Aztec records before the arrival of Europeans in the 1500s (Desowitz 1991). Other references support the theory that Europeans introduced malaria to the Americas (Schlagenhauf 2004; Dubisch 1985; Carter 2002).

In a contrasting opinion, Merrens and Terry, writing in their paper, "Dying in Paradise," state that South Carolina was only perceived as being healthy before 1750. This false sense of well-being was largely due to propaganda generated to encourage emigration to the colonies. Merrens and Terry give an example from the 1600s. Emigrants were being told that if they came to Carolina, they could expect a long, healthy life and wealth from their labors. William Hilton, for whom Hilton Head Island is named, wrote in 1664 that the land was so healthy that one could sleep outside naked and not become ill. Hilton said also that the natives were healthy and that many lived to old age. During the seventeenth century, it was common to attribute health and productivity to the areas near latitude 35 degrees, such as South Carolina, Antioch, Judea, and the most productive regions of China. In fact, Sir Walter Raleigh is quoted as saying that latitude 35 degrees is equated to Paradise. Merrens and Terry suggest that a legend was created about an ideal environment and that immigrants continued to come to Carolina even after current residents realized they were not living in paradise (Merrens and Terry 1984).

Henry Savage of Camden, South Carolina, in his book *River of the Carolinas: The Santee*, refers to a description of the Carolinas by Robert Horne in 1666: *there is no Plantation anywhere the English have been that is so good as Carolina.* Horne's writing appears to be enticing those living in Barbados to emigrate to a better place: *Carolina has ground as rich as any in the world, clay for bricks, lime for building, woods, groves, marshes and meadows. There is little underbrush and travel is easy by horseback or foot. There is a wide variety of trees, fruits, and medicinal plants. The woods are full of deer and turkey and all taste better than in England. The meadows are good for cattle and any grain sown will thrive. The rivers are filled with an abundance of fish. Ducks and geese are plentiful in winter. The summers are not too hot and the winter short and moderate, best agreeing with the English Constitutions . . . If therefore any industrious and ingenious persons shall be willing to partake of the felicities of this Country, let them embrace the first opportunity, that they may obtain the greater advantages . . . If any Maid or single Woman have a desire to go over, they will think themselves in the Golden Age, when Men paid a Dowry for their Wives; for if they be but Civil, and under 50 years of Age, some honest Man or other, will purchase them for their Wives* (Savage 1968, 47).

Between 1700 and 1750, many planters began to make the association that a high proportion of deaths occurred during the summer months and that malaria was the cause. In Christ Church Parish Mt. Pleasant, one-third of all females died before age fifty. Sixty-nine percent of male deaths occurred before the age of five. Of this 69 percent, half died during the peak malaria period from August to November. As an example, Merrens and Terry mention Henry Ravenel (1729–1785) and Mary de St. Julien (d. 1779) of Hanover Plantation. Married in 1750, they had sixteen children, only six of whom lived to adulthood. Of their seven daughters, none lived to age twenty. Peter Gaillard and his wife, Elizabeth Gourdine Gaillard, had twelve children; only two lived past age twenty-three.

By 1750, the residents of St. John's recognized they were not living in Paradise, even if it was latitude 35 degrees. The perception of the plantation as a healthy place to live began to change. Perceptions abroad were changing also. A saying in England in 1750 was: *They who want to die quickly, go to Carolina* (Merrens and Terry 1984, 549). A German traveler reported in 1737 that *Carolina is in the spring a paradise, in the summer a hell, and in the fall, a hospital* (Merrens and Terry 1984, 549).

Then, as now, when there is no explanation for natural disaster, some religious groups blame sin. Rev. Robert Maule, the first minister to St. John's Parish, contracted a fever in 1709 and seven years later died of *Consumption and a lingering fevere*. During his illness, he reported back to his superior, the Secretary of the Propagation of the Gospel in Foreign Parts, Church of England, that there was a high rate of mortality *for those Things the Lord does yearly visit, sending pestilential diseases among Men and Beasts, which yearly sweep away Numbers of both . . . was due to infidelity, heresy, and other breaches of morality present in Christ Church Parish* (Merrens and Terry 1984, 541).

Despite continuing interest and concern with healthy living, the fact was that the cause of malaria was unknown. With the proliferation of farmland and rice fields in St. John's, the number of sites for stagnant water increased. It was not recognized until later that this produced more breeding grounds for mosquitoes and the spread of malaria. Ironically, even physicians and scientists seeking cures were not immune to the ravages of this disease. Dr. Francis Peyre Porcher, renowned botanist of St. John's, lost his son Leigh to malaria in 1865: [Dr. Porcher] *devoted years to the study of malarial and yellow fevers in his effort to find the cause and cure—but the mosquito was not suspected then and no effort made to drain a pond directly behind our home in Pinopolis—and my brother, Leigh, fell a victim to country fever when he was nine years old. His death was a calamity* (W. Porcher).

Thirty-two years after Leigh's death, the mosquito was identified as the carrier of the malarial parasite. It is now known that the female anopheles mosquito is the carrier of the infectious malaria organisms, *Plasmodium falciparum* and *Plasmodium vivax*, which live as parasites in red blood cells.

Drainage ditches had been dug in old Pinopolis, but it is unclear whether they were dug because of a malaria outbreak or to prevent one. In March 1852, Thomas P. Ravenel recorded in his journal that Pinopolis villagers were to begin digging ditches and that an old ditch on the west side was opened. In April 1852, Ravenel noted in his journal that villagers were to dig a ditch from the head of Turkey Bay to Bunker Hill. Also, the ditch dug the previous year was opened and deepened. Five years later, in April 1857, Ravenel noted that villagers were again opening and cleaning ditches in and around Pinopolis.

I can distinctly remember playing in a ditch near Charles Cain's house at least sixty years ago. Unaware of any useful purpose for it at that time, I now believe that it was one of the original drainage ditches. In March of 2006, I went in search of that ditch. One can easily see its remains today. On the west side of Pinopolis road, just north of the Massalon house, there is a small ditch overgrown with trees. Its westward extension has been filled in by a housing development. (See Pinopolis Peninsula map.)

chapter II

Pineland Villages

All along the ridge of Pineland are found the little summer villages . . . in fact, they are scarcely villages—hamlets rather, composed usually of from ten to fifty houses scattered about with their outbuildings among the pine trees.

Rev. Dr. Robert Wilson

Please see map on page 191 for locations of villages described in this chapter.

THE CLASSIC PINELAND VILLAGE OF THE SOUTH Carolina Low Country was defined by the semiannual migration of planters and their households from the plantation to the village in May and back to the plantation in November. Summerville and Walterboro, two of the earlier pineland villages of the South Carolina Low Country, were settled in the late 1700s. Early on, Summerville and Walterboro had year-round residents and developed into incorporated towns. Patterns of growth and development that occurred in those towns did not occur in the pineland villages of St. John's Berkeley.

While this book focuses on Pinopolis, other well-known villages of Berkeley County were Pineville, Raccoon Hall, Whitesville, Cordesville, the Barrows, New Hope, Spring Hill, Honey Hill, and Gravel Hill. A less-known settlement, Woodville, was described as being east of Pooshee.

In his book, *Half Forgotten By-ways of the Old South*, the Rev. Dr. Robert Wilson observed: *All along the ridge of Pineland are found the little summer villages . . . in fact, they are scarcely villages—hamlets rather, composed usually of from ten to fifty houses scattered about with their outbuildings among the pine trees. It is a strange reversal of social laws where the peasantry live upon the manor and the lords of the land group themselves into hamlets . . . these villages are perfectly unique. There are no streets; only broad roads meandering uncertainly about among the enclosures . . . The most striking feature is the long, broad piazza, with its benches and oaken chairs. No gardens are allowed, as cultivation invariably makes the place unhealthy, but the yards are all enclosed, and everything looks fresh with the constant white-washing* (Wilson 1928).

The one aspect of plantation/pineland village life that contrasted with the lifestyle of those not of the planter class was the semiannual move. Children and women generally did not welcome leaving the plantation. Reflecting upon his boyhood years spent in Pineville about 1815, F. A. Porcher noted: *This was a change* [the move from the plantation to the village] *that none of us much enjoyed; for plantation life has great charms for a child and this was now exchanged for the dullness of* [the pineland].

By 1844, Porcher had established himself as a planter at Somerton Plantation and was following the same semiannual migration pattern as his fellow planters. In a poignant description of living in Pinopolis, he said: *There was no church in the village, but divine services was performed almost every Sunday at Whiteville, which was only four miles below, so that every one who desired it could go to Church. One day in the week Mr. Wallace . . . would perform services in one of the houses of the village, and he always had a full congregation. I used to think this the bitterest commentary upon the dreariness of pineland life; the people having no sort of amusement or recreation would be glad even to go to Church. In truth, except those to*

Moving day at Somerset Plantation. *Courtesy of the Cain file, South Carolina Historical Society.*

the manner born, a pineland life is a dreary life to a woman. There is no inducement to [leave one's property], *a ride or a drive would only carry her among more pine trees. The walks have nothing to tempt one out of the house. Those who are used to it* [village life] *are perfectly content, but it is the contentment of ignorance. They have never known anything better; and of course cannot be very much dissatisfied with what they have all their lives been accustomed to. Yet I have seen some Charleston women who appeared to fall easily into the train of life of a pineland village and seemed perfectly content with it. Possibly they only assumed the virtue of contentment, whilst at home they gave free expression to their real feelings* (F. Porcher 1943).

One of the disappointing aspects of moving from the plantation in May was that it coincided with the most beautiful time of year. Flowers were blooming, trees were bursting with bright green leaves, grass was greening, fields were freshly plowed and planted, and farm animals and wildlife were more active. William Cain remembers growing up at Somerset in the early 1900s: *Incredible as it may sound, we moved from the plantation each year in mid-May and then returned in mid-November . . . Somehow, the time spent in the village always seemed like a temporary interlude, while that spent on the plantation was real living* (W. Cain).

Moving day was a big event with wagons, mules, and crates and boxes filled with furniture, kitchen utensils, clothing, food, and books and other items for entertainment. Anne Simons Deas (1845–1928) grew up on Coming Tee and Buck Hall plantations on the Cooper River and spent her childhood summers at the pineland village of Cordesville. In later years, she wrote: *At last the move is over . . . it seems such a pity to leave the plantation, where everything is more beautiful than ever in the full glory of foliage, and come to this monotonous-looking pine barren, where the trees stand so thickly that there is practically no horizon. And as for the moving—I don't know how anybody can stand it semi-annually. We breakfasted much earlier than usual, but before we sat down to table, the big six-ox-wagon came creaking into the yard . . . The mule-wagons came up to the house, and I flew to pack my own belongings . . . very little of the heavy furniture was being moved . . . an out house disgorged plain bedsteads and substantial cane-seated chairs, an old dining table of lighter build, and a huge box of crockery which belongs to the summer life, and is never unpacked in winter. The house proper contributed trunks, boxes, mattresses, small tables, a quantity of pantry ware, two wire safes, and a goodly array of storeroom belongings . . . A one-horse cart . . . was backed up to the kitchen steps and loaded with pots and pans and Dutch-ovens and spiders, a large marble mortar, and numberless pails and piggins.*

Meanwhile, the packing of the big ox-wagon went gaily on. The straw-frame was on it, and its capacity seemed practically unlimited . . . There were boxes, small chests, little presses, baskets of every shape, size, and kind, pots and pans, wash-tubs, and bundles of bedding; while long coops of fowls and lots of children of all sizes occupied [space in the wagon]. *The children were clad in festive attire, and were furnished with large hunks of corn bread, wherewith to beguile the tedium of the journey. It took longer than you would think to get all this done.*

The wagons left first, and Anne and her family stayed on the plantation until after lunch. She rode the five miles to the village in a carriage. *When we drove up to the door* [of the village home], *the wagons had just been unloaded, and confusion reigned . . . inside the house . . . women servants were rapidly reducing chaos to order.* In the evening, she concluded: *I sat on the front steps and watched the last rays of the sun gilding the tops of the great pines before us* (Deas 1910).

In *Summer Migrations and Resorts of South Carolina Low-Country Planters*, Lawrence Brewster described a typical moving day. *These little villages, completely deserted in winter . . . came alive early in May. They were awaked one morning by the advance guard from the plantations, sent to prepare for the arrival of the planter families . . . The dwelling house was cleaned and tidied; its simple furniture, consisting of pine tables, oak chairs, and benches, and beds of the 'bed-and boards' variety, was dusted and repaired and augmented by cast-off sideboards and wardrobes and the 'indispensable piano' brought from the plantation. The cottage and its few outbuildings—kitchen-house, slatted stable and open vehicle shed, cow pen and padlocked hencoop—were freshly whitewashed; the yard, where no cultivation was allowed, was swept clean. In the fire-stands—square, earth-filled frames mounted on posts, which stood in the front yard—were placed piles of lightwood knots and pine straw. Thus aroused from its hibernation, the village was soon ready to receive its summer residents* (Brewster 1947).

William Cain remembered the move from Somerset to Pinopolis in this way: *Since Somerset was only two and one half miles away, moving was not nearly as burdensome to us as to our more distant neighbors like the Lucases at White Hall or the Porchers at North Hampton, each five or more miles from Pinopolis. Even at best, however, the whole affair was disrupting and frequently objects would be totally lost, only to be rediscovered when the next move was made in the opposite direction* (W. Cain).

Once the household was firmly settled in the summer home, a well-defined pattern of activity emerged. Frederick A. Porcher, who lived first in Pineville and later in Pinopolis, described the daily routine in his memoirs: *The habits of every house were alike. At sunrise breakfast was served, and the planters went out to visit their plantations . . . the planter was generally on his horse . . . hunting also afforded the means of passing the time . . . after the morning's ride was over, the post office or village store was the general rendezvous and lounging place. Here politics and crops were the never-failing topics of conversation.*

At one o'clock dinner was served. A portion of the afternoon was always devoted to sleep. Every piazza was furnished with long benches, and these formed the rude beds on which the gentlemen invariably indulged in the luxury of a siesta . . . the siesta over . . . hot tea and cakes would be offered. After supper was the time for visiting. *The social life commenced . . . Visits were unceremonious . . . young ladies usually dressed in white; the aged were clad in graver colors. The guests were received in the piazza. No one expected to be invited into the house, and persons might spend a season in social intercourse with the people, without seeing the interior of any house but their own. No refreshments were offered or expected. But though the visiting was done at night, and the piazza the reception room, the company did not sit in the dark. Every pineland summer residence has in front of it,* [in the yard] *sometimes both in front and rear, a rude staging deeply covered with earth, which is technically called a lamp. On this stage, as soon as the shades of evening begin to lose themselves in the night, knots of lightwood are burned, and the fire continues to be kept up until the family goes to bed. Several objects are affected by this fire; it gives light to the dwelling and dispels the deep gloom of a pineland night, and the superior brightness of the outer lamp attracts from the house those multitudinous insects which swarm in these regions, and which would otherwise be attracted by light of the dwelling . . . the material burnt was the dead pine leaves and cones . . . the numerous pines in the villages furnished these in great abundance . . . and it was easy for a man in a short time to collect as much as would serve for the evening lamp . . . These evening yard fires always rise before me as my memory wanders back to the scenes of the past . . . know that their lurid light in the intense gloom of night gave a weird-like aspect to the surrounding scenery. A fire too has always a strange fascination for a child. We would sit around it, and when the flame would be made bright by stirring it up, or throwing fresh fuel around it, we would delight in showing our audacity or our agility by leaping through the fiery vapour. With fires burning before each dwelling a cheerful tone was diffused over the whole village; without it the gloom and darkness were intense . . . By ten o'clock social life was over . . . Whilst the visitor was preparing to return home, the servant lit his lantern and with this simple but necessary escort, she trod the streets of the village with as much security as the halls of her own mansion* (F. Porcher 1943).

Brewster's description of village life is livelier: *The* [village] *roads were much traveled by individuals and families on foot, on horseback, and in carriages, passing back and forth from cottage to cottage. Every piazza was the scene of much gaiety and hospitality . . . For organized evening entertainment, the summer residents had cottage card parties, musicales, and dances; occasionally there were more formal and public assemblies and balls* (Brewster 1947).

The Rev. Robert Wilson, MD, of Charleston, had family connections in St. John's and Pinopolis. He was also a frequent guest preacher at Trinity Episcopal Church in Pinopolis. In 1875, Reverend Wilson wrote concerning village life: *Everybody knew everybody else, and, having little else to do, went to see everybody else every day and at all hours. Sociability became almost oppressive until one got used to it . . . The routine never altered, but nobody ever tired of it.*

Wilson, like F. A. Porcher, particularly enjoyed the evening yard fire. *These fires were kept burning until bed-time, serving the double purpose of furnishing light and of attracting and destroying the swarms of mosquitoes, which would otherwise have murdered sleep. One never forgets the wild and weird effect of such a scene. The straight, shadowy trunks of the ancestral pines now fading into darkness, now sharply defined in the ever-varying light, the cloudy foliage, distinct in its ghostly uncertainty of outline, mingling like the dissolving views of the phantasmagoria with the dense columns of smoke* (Wilson 1928).

Whether in Pinopolis, Cordesville, or Pineville, the evening fires gave peace, pleasure, and comfort to the village people for a hundred years as well as providing a defense against the nightly invasion from insects. However, the fear of uncontrolled fire was always present. Before leaving the village in the fall, all residents raked and burned the pine straw from around their homes and outbuildings to lessen the chance that a forest fire would reach the wooden structures.

The destruction rendered by fire is difficult for us today to imagine. Fire posed a real threat throughout the late 1800s, as can be observed in the *News and Courier* of Charleston on December 2 and 3, 1886: *On December 1, 1886, Berkeley County was swept by a devastating fire from St. Stephen to the Charleston County line on both sides of the North Eastern Railroad, including the section around Stoney Landing. There was great destruction of turpentine trees. It was reported that the village of Pinopolis was barely saved from destruction. There, Mrs. Kate Porcher lost her dwelling and all her outbuildings. Trial Justice Cain lost his outbuilding, and only hard work saved his dwelling. H. H. Winter lost his barn and his entire harvested crop. Many Negroes around Pinopolis and Moncks Corner lost all they possessed. There was considerable loss of property and provisions around the village of Oakley.*

Thomas P. Ravenel, a Pinopolis resident, described this 1886 fire in his journal: *There was great danger that the whole village would be burned as many wells gave out. The smoke was so dense that men would almost ride over persons before seeing them & when Mrs. P's* [Kate Porcher] *house was burned, those a hundred yards did not see the fire. Those who were there describe it as a day of horror & say they would rather go through another earthquake than such a reign of fire* (T. Ravenel).

From 1899 to 1910, John Henry Porcher, who was living in Pinopolis and at various plantations, recorded in his journal the location of the following fires: Bossis Plantation; Peter Gaillard's residence in Eutawville; and Hattie Wilson's stable (J. Porcher).

Thomas P. Ravenel reported other fires in his journal from 1867 to 1894: his own home at Woodboo; a barn and stable at Belle Isle; a house on Tibbicudlaw; the barn at Moorefield; the house at Bluefield; the house on Wilton Plantation; two gin houses at Northampton over a period of three years; a dwelling on Poplar Hill; the gin house at Somerset; the house of Dr. Sidney Kirk; the house at Chapel Hill; J. St. Claire White's store in Moncks Corner; and the Pinopolis house of Mrs. Charles Macbeth (T. Ravenel).

William Cain remembered the attraction and destruction caused by fires: *Forest fires . . . were frequent and often terrifying. The tenants thought they decreased the boll weevils. They* [the fires] *spoke of wild beauty but were very destructive. I remember one that burned five thousand acres of our land alone in one twenty-four-hour period, and fighting it produced in me the greatest fatigue I have even known. For some reason the smell of burning wood still takes me back to plantation days more surely than anything else* (W. Cain).

Pineville Chapel was built in 1810 and restored in 1940.

The rituals of plantation life, with the traditional annual move to the pineland village changed drastically after the Civil War ended in 1865. Most plantation owners were financially devastated. Those who could afford to keep their plantations continued the annual migration. Others lived permanently in the villages; some were forced to find means of support elsewhere.

Pineville

Pineville, the first pineland village in Berkeley County, was settled in 1794 by Captain James Palmer, Peter Gaillard, John Cordes, Phillip Porcher, Samuel Porcher, and Peter Porcher (DuBose 1887). Pineville is in St. Stephen's Parish.

At its point of greatest prosperity, Pineville consisted of sixty well-built houses on one- to two-acre lots. Because pine trees were considered essential to a healthy environment, Pineville residents could be fined five dollars for the cutting of trees. In 1791, before anyone had thought about a pineland village, planters in the area had laid out a racecourse, and by 1792, their Jockey Club had ninety-seven members. As Pineville grew, a public library was organized in 1810, and by 1826, a larger building was needed. There was also a ballroom for elaborate cotillions. The Pineville Chapel, which is still standing, was built in 1810 and restored in 1940.

There can be little doubt that no other villages ever equaled Pineville in its culture and education. A grammar school, the Pineville Academy, was established in 1805. Its reputation was so highly regarded that students went there from all over South Carolina. Men shared a high degree of intellectual curiosity and placed a priority on education. Many of the planters and their sons were schooled at Harvard, Yale, Princeton, and in Europe.

F. A. Porcher commented upon the men of Pineville who were his contemporaries: *If there was ever on earth a community that really deserved to be called a happy one, it was that society which revolved about Pineville as its centre of attraction and of cohesion. In point of social status perfect equality prevailed; as regard fortune little disparity, but even those who were least favoured, had more than their wants demanded. As for their manners such perfect simplicity prevailed, that there was not only no inducement to extravagance but it would have been severely frowned down. The occupation of all was the same, all were tillers of the earth and planters of cotton so that all sympathized in the pursuits of each. An elevated tone of moral feeling characterized the society almost perfectly free from any tinge of religion. The society had also the refinement of intellectual culture, there were few men who had not been educated at a college; but I am bound in candour to add that except from the general tone of conversation few strangers would have suspected that so many of the people had received such a high grade of mental culture* (F. Porcher 1943, 80).

Only one of the early homes remains. The South Carolina Inventory of Historic Places records this home as having been built around 1820 by a Gourdin. Tradition has it that the old Pineville Post Office, dating from 1830, and the Pineville Library building, dating from 1826, are the structures now on the north side of the Gourdin house. David Hoffman, who now owns the home, has many of the original library books.

In 1832, the U.S. Federal Census in Pineville recorded a total population of 789; 235 were white and 554 were black. Pineville suffered fever epidemics in 1833, 1835, and 1836, which caused a mass exodus from Pineville to other villages. Some residents later returned, but Pineville never enjoyed its former prestige. In 1865, Major Rogers of the Union Army under General Potter's command burned most of the homes.

Raccoon Hall and Jack's Hole

Raccoon Hall was located about five miles west-southwest of Russellville and close to Jack's Hole, a spring on Pooshee Plantation. Henry William Ravenel (1814–1887) recorded in his journal: *My father erected a rough temporary house at "Raccoon Hall" North of Pooshee, where he lived two or three years. Several families then determined to make a summer settlement at "Jack's Hole," a high ridge east of Pooshee . . . In [18]38 I moved to Jack's Hole village where I resided until '43, when I removed to Pinopolis, there I lived until '53 when I removed to Aiken* (H. Ravenel 1947).

When the remaining homes in Raccoon Hall and Jack's Hole were abandoned after the Civil War, it is thought that freed slaves then occupied them, and the settlement took on the new name of Coon Hill. This area is now under Lake Moultrie (E. Bull 1965).

The Barrows, also called Raccoon Hall

In *A Day on Cooper River,* John Irving, MD, states that residents of the plantations along the Cooper River in Lower St. John's Berkeley sought refuge from malaria at Cordesville, the Barrows, and Whitesville (Irving 1969).

Raccoon Hall was the name initially given to a settlement south of Moncks Corner. It later became known as the Barrows when numerous mounds, which were determined to be Native American burial sites, were discovered there. Barrow is a Middle English word referring to a burial site where the body is piled upon with earth instead of being buried.

In the early 1800s, planters from the Cooper River built summer homes at the Barrows. The Elias Bull papers at the South Carolina Historical Society contain a rough map of the Barrows showing old homes in relationship to present-day Oakley Road and Highway 17-A.

In the *Charleston Daily Courier* on March 9, 1862, an anonymous journalist noted: *I have passed a very agreeable day at the Barrows. About 15 families can be considered regular residents of the Barrows . . . Col. James Ferguson; Dr. Barker, Senator and provost* [ca. 1828]; *Thomas Broughton; Thomas Milliken; Rev. A. F. Olmstead; Dr. C. B. Lucas; Peter Nelson; James Poyas; Catherine G. White; Peter Gaillard Stoney; Dr. Charles Edward Tennent; John Porcher; Dr. Motte; Poppenheims; Carsons; Stevens; and Heyward . . . It is thought that Col. Ferguson built a home here around 1842 and was a prominent resident . . . The venerable Col. James Ferguson, one of the few specimens remaining to us of a Carolina gentleman of the old school, is in this favored community, properly regarded in the light of a Patriarch . . . everyone in the village seems to look up to him and to love him . . . A very neat Episcopal church has recently been erected, capable of accommodating all the piously disposed in the neighborhood . . . The Rev. A. F. Olsmstead officiates every alternate Sunday.* Some of these planters and their families are also mentioned in *Names in South Carolina* (1965).

The Episcopal Church of the Holy Family in 1970. *Courtesy of the Elias Bull Collection, Charleston County Library.*

The Barrows' first settler may have been Dr. (and Senator) Henry Barker in 1828. It is thought, however, that the overseer employed by Mrs. Motte and

Mr. Thomas Broughton may have had a home here earlier. Planters of Lower St. John's on the Cooper River resided at the Barrows until the end of the Civil War.

Some time in the 1960s, Panchita Heyward Grimball, whose parents owned a Cooper River plantation, wrote her memoirs of life in the Barrows: *I now come to the village in which I spent the summers of my early life . . . The Barrows was about one mile from a community called "Burbage Town," which was settled by country people, who did not own slaves, and were very, very poor . . . Mr. Steven De Veaux of Pinopolis started a mission among these people and worked faithfully until his death in 1905 . . . The Stoney family moved to Pinopolis when I was quite young as Mr. Stoney* [Samuel Porcher Stoney] *had bought a plantation quite near that village. We were really distressed to see them go, for we were dear friends.* After the Civil War, most of the homes in the Barrows were abandoned and as time passed, many rotted or burned. A timber company later bought most of the land known as the Barrows (Grimball).

Also in the memoirs of Panchita Heyward Grimball is her account of a church in the Barrows called Guild Hall, now known as the Church of the Holy Family. She doesn't give a date of origin but states that Mrs. Motte gave a small piece of land for a chapel, which seated about sixty people. The Rev. A. F. Olmstead first held services at the Barrows in 1863. Grimball recalled beautiful wide pine floorboards in the chapel. She said that after most of the residents moved away, the chapel was taken down and rebuilt where the present Episcopal Church of the Holy Family now stands. The Church of the Holy Family building was constructed around 1942. It has been a struggling mission for most of its existence. In September of 2005, it was closed temporarily and the congregation invited to participate at Trinity Church, Pinopolis.

Whitesville

Whitesville was settled by John Sims White (1820–1861), who owned Gippy Plantation. The village, about two miles north of the Barrows, was never a large community, but it did have a school. In 2005, that school building was moved to its present location next to the Berkeley Museum and Heritage Center in Moncks Corner, where it is being restored. Whitesville had a chapel before Pinopolis or the Barrows. In 1836, the Rev. Edward Thomas began service in the parish of Biggin Church. He was also responsible for services in the chapels at Whitesville, Cordesville, and Strawberry (A. Thomas 1957). The Bible at the Whitesville Chapel was given to Trinity Episcopal Church in Pinopolis. No remnant of the village of Whitesville remains. *The trees have all been cut down and now large fields of cotton and corn greet the eye, with never a hint of where even the chapel or the Rectory once stood* (Grimball).

Cordesville

Cordesville was settled by planters of the upper Cooper River. Records show this pineland village in existence before 1819 (E. Bull 1965). Families who lived there were Cordes, Ball, Harleston, Prioleau, White, Fishburne, and Gibbs (Grimball). Cordesville had a small church in the 1850s and even a bowling alley (Deas 1910). Although most of the old homes are gone, Cordesville is still a residential community.

The Episcopal Church of the Holy Family, the Barrows, 2005.

chapter III

Pinopolis and Its People, 1834–1865

Stories have to be told or they die,
and when they die, we can't remember
who we are or why we're here.

Sue Monk Kidd, *The Secret Life of Bees*

SEVERAL CONSECUTIVE "SICKLY SEASONS" IN PINEville caused its planters to look elsewhere for a summer retreat. In his journal in October 1861, Henry William Ravenel recalled: *It was the fatal summer of 1834 that destroyed confidence in the health of the village* [Pineville] *& caused a large portion of the inhabitants to remain away. Some afterwards returned, but many have permanently elsewhere fixed themselves . . . The summer of 1834 was the last entire summer I lived in Pineville.* Ravenel lived in Jack's Hole and then moved to Pinopolis in 1838 (H. Ravenel 1947).

In 1834, Frederick Adolphus Porcher (1809–1888) and Dr. Morton A. Waring (1809–1875) initiated the settlement of Pinopolis, this new village in the pineland. Porcher described in his memoirs how Pinopolis came to be: *Dr. Waring* [Morton Alexander Waring] *and I had determined to build houses for ourselves on a pineland west of Begin Swamp and about four miles south of Somerton . . . By agreement we joined our forces and directed all our efforts to finish one house, which we agreed was to be taken by Dr. Waring, as he had no summer house, and I had Pineville to retreat to. By the middle of June* [1834], *my house was so far completed as to be habitable. It was a log house . . . about forty feet long on the north and south sides, and two pens at the ends, made of logs twenty feet long about fifteen feet apart, so that the house when*

Dr. Morton Alexander Waring, one of the first two settlers in Pinopolis, built the first home in Pinopolis in 1834. *Courtesy of Waring Historical Library.*

finished and roofed in had three rooms . . . But I must say that I do not think it creditable to our civilization that persons brought up as we had all been, should have been satisfied with such houses even as temporary residences (F. Porcher 1943).

F. A. Porcher is credited with giving Pinopolis its name. *I dated for the sake of the jest several letters which I had occasion to write, Pinopolis, and this became the name of the village which now numbers between twenty and thirty excellent dwellings* (F. Porcher 1943, 154–155).

Thomas Porcher of White Hall wrote to his brother, Dr. Peter Porcher, in 1837: *In consequence of having so much to do I have rented out our house in Charleston and have removed the house which young Mr. Marion built on a part of Goshen to the pineland of Dr. Waring and Mr. Cain, where I expect to live until July and then go to the back country . . . Pineville will be almost deserted this summer . . . Charles DuBose is also building there* [Pinopolis], *and F. A. Porcher and Dr. John Porcher are also to be there* (P. Porcher).

According to F. A. Porcher, these families lived in Pinopolis in 1844: Dr. Morton A. Waring and family; Porcher's uncle Isaac Porcher; Henry F. Porcher; his brother, Dr. Peter Porcher; Mrs. Gaillard; Mr. Cain and his family; Mrs. William Porcher and her sons and daughters; Thomas Walter Peyre, living with Henry William Ravenel; and Mrs. Margaret Evance Cantey [widow of James Cantey whom she survived by almost sixty years] with her grandchildren, Susan and Charles DuBose, and Charles's wife and children (F. Porcher 1943, 101–103).

F. A. Porcher grew up on Cedar Spring Plantation, which his father, George Porcher, owned. F. A. Porcher spent his childhood summers in Pineville and attended the Pineville Academy. He spent a year at Partridge's Academy in Norwich, Connecticut, where one of his classmates was Morton A. Waring. Porcher graduated from Yale University at age nineteen. Not feeling drawn to a particular profession, he tried reading law in Charleston but found it unsuitable to his temperament.

Porcher felt that he was supposed to be a planter and purchased Somerton Plantation in 1831. *It seemed to be taken for granted that I was to continue to be interested in planting even if I were a lawyer . . . My residence at Cedar Spring the last winter had fostered the wish to join the fraternity of planters, though I was conscious that I had no vocation for the business . . . Samuel J. Palmer, who was the owner of Somerton, was desirous of selling; he offered it to me, and I did not long hesitate before I accepted the offer. It was an unwise act on my part* (F. Porcher). Porcher was most unsuccessful at planting and recognized that he was never suited to be a planter. *I was not intended for a planter . . . it was not from the want of intelligence; my knowledge of the planter's business and profession was equal to that possessed by my neighbors; but it was in the practical application of my knowledge that I failed* (F. Porcher 1943, 83). Porcher struggled as a planter until 1848 when he accepted a position as professor of "belle lettres" at the College of Charleston. In his later years, he wrote extensively about his life as a member of the Low Country planter society.

A keen observer of human experiences, Porcher described an interesting contrast between a planter born into the class and a successful self-made planter. During the summer of 1839, F. A. Porcher traveled to Greenville, South Carolina; Asheville and Warm Springs, North Carolina; and several other resorts. He described meeting Benjamin Allston of Georgetown, who spent his summers in the same areas that Porcher was visiting. Porcher wrote, *I felt painfully conscious that I was out of my place, for I was not so much a planter as a man living upon the proceeds of planting . . . One of the distinguished habitués of these regions . . . was Mr. Benjamin Allston of Georgetown . . . He had been a very successful man, commencing life I believe as an overseer. By means of industry and thrift, he had become one of the richest rice planters on the Waccamaw . . . He was a keen judge of character and knew how to distinguish between affectation and pretension, and genuine merit. I always thought better of myself because Mr. Allston seemed inclined to cultivate my acquaintance. His conversation was that of an utterly uneducated man . . . One night a traveling phrenologist gave us a lecture upon the new science; he illustrated afterwards by undertaking to read characters by the phrenological developments. Mr. Allston was a patient assistant during the lecture, and he looked curiously upon the professor as he was manipulating the heads of his subjects and pronouncing their characters. At last he* [Mr. Allston] *said in tones which were heard distinctly all over the room: well, I h'aint seen sich a sight since I was a little shaver, when my Mammy used to sarch my head for creepers. He* [Allston] *was precise and accurate in everything he did or said* (F. Porcher 1943, 90–92). In spite of their cultural differences, Porcher and Allston maintained an ongoing friendship.

Porcher describes his impression of the general planter's attitude toward church and religion during his boyhood years, 1815–1820, on Cedar Spring Plantation and in Pineville: *I have said that the Society had a high moral tone untinged with a religious sentiment. I do not wish to be understood as intending to deny the existence of the religious principle, but I think it is true that the religious feeling was utterly unknown; and I am very sure that the puritanical development of the religious idea would have been utterly unintelligible when I was a boy. People went to Church; but the idea of a puritanical observance of Sunday would have been regarded as somewhat bordering on insanity; thus it never occurred to me that my school lessons should be prepared on Saturday instead of Sunday evening; on the contrary, Sunday evening was always devoted to study, and I am sure this was much better, even in a moral point of view, than to have the whole family trying in vain to get comfortably over the many hours which, by the puritanical custom, compose a dreary Sunday. The children were not neglected in point of religious teaching; we were all taught the catechism, and I have no doubt but that we were better instructed in religious learning than those are now who have all the boasted advantage of Sunday schools and Bible classes; but though the intellect was religiously trained, there were no efforts made to cultivate the religious affections. Sunday was notoriously a day for traveling. If one was desirous of spending a week in town, he went on Sunday and returned on Sunday. For a very long time, the church had no male communicants; and I must add that profane language was common among the men, and very much so among the boys. I remember when I was the only boy at school who did not indulge in such language; from this I was kept by the sedulous care of my mother; and not having contracted it in boyhood, I never practiced it as a man* (F. Porcher 1943, 85).

Porcher describes his father as an indifferent planter who had no wealth. He was not aware of it as a child but later remembered his mother's attempt to free their estate from debt. He said the family at home never even drank tea unless they had guests. They bought used furniture and painted over nicks and scars in hopes that visitors would not notice. However, the Porchers spared no expense when it came to having a huge coach, drawn by four horses, when visiting neighbors like the Broughtons, Ravenels, and Marions.

He says, *The visitors came early; they generally dined earlier than usual in order to have a longer afternoon, and it was a point of honour to give your visitors their tea early that they might reach home before it was dark . . . It happened that few of our neighbours had less than four or five miles to ride, and those roads were not remarkable for their smoothness; in fact the crowd of wagons bearing country produce to Charleston was so great that the public roads were often almost impassable, and at every hundred yards you would come across some place where a wagon had stuck and see the primitive leverage by which it had been extricated.*

Porcher was well aware of class differences. He associated a certain style of dress with a certain class: *Our society was unique. Consisting entirely of planters' families, all of whom occupied the same social position, there was a total absence of affectation or pretension. I looked with confidence upon these men* [patriarchal planters like the Ravenels, Duboses, and Mazycks], *whose blue cloth coats, knee breeches, high boots with tassels hanging down the front, distinguished them from the people of the parish in humbler walk of life and the queue* [ponytail] *which gently flapped the collar of the coat gave them an appearance of respectability in strong contrast with the closely cropped hair of the other people . . .* [By contrast,] *the Wagoners looked so unlike the men I was accustomed to see, travel stained, rough, uncouth, I never could approach them but with dread; and in my rides* [as a teenager living on Cedar Spring plantation] *one of the most unpleasant sights that could present itself was the mysterious white covered wagon, slowly and toilsomely plodding its weary way to the metropolis.*

As a young man, Porcher had a high regard for the medical profession as it was one of the few socially acceptable professions other than planting. During the illness of his first wife, Rebecca, Porcher witnessed medical incompetence and quackery that changed his opinion.

In 1832, Rebecca gave birth to premature twins who died; in 1833, another child was stillborn. In 1833 or 1834, Porcher then took Rebecca on an extended trip to the "back country." During this trip, they stopped in Winnsboro, South Carolina. *In the dead of night my wife called to me that she had a hemorrhage. I was extremely frightened and knew not what to do. It proved however to be very slight* [and] *stopped . . . Whilst she was dressing the next morning, it returned.* A physician was called and immediately bled her. *I have since that time had many sad opportunities of becoming acquainted with this malady* [uterine bleeding], *and my knowledge subsequently gained now directs my judgment of the treatment she received. But at that time I was happily utterly ignorant. The hemorrhage was neither profuse nor alarming; it might have been controlled or at least checked by styptics. The revulsion of feeling in me was so great that I fainted when the Doctor was bleeding her . . . the bleeding had stopped.*

Frederick A. Porcher built the second home in Pinopolis in 1834 and gave the village its name. *Courtesy of Richard D. Porcher, PhD.*

She was forbidden to have meat or stimulants and could have only vegetables. *I never suspected what I know now, that he was fixing the disorder permanently upon her by depriving her of the means of recovering strength; but he practiced according to the light of his times, and she is not the last that I have had the pain of seeing die under the same ignorant practice. I was happily entirely ignorant both of the extent of her danger and of the mistakes which were committed . . . Oh! How sad it is to see when you have imprudently broken the stern rule of the physician and persuaded the patient to eat a little piece of meat, how sad I say, to witness the smile of content which then beams upon the sufferer's face, and to be made to believe that she must suffer for that little indulgence.*

They returned to Charleston. *Doctor Glover saw her and prescribed as much meat as she could eat, but his prescription came too late; in less than ten days after her arrival in town, she died* [November 1834] (F. Porcher 1943).

Porcher's confidence in the medical profession continued to erode. Following his wife's death, he traveled in Europe from June 1835 to November 1836. In Paris, he encountered Charlestonian Mrs. Ellinor Gourdin McBride and her daughter, Elizabeth, who had been sent to Europe for her health. Porcher thought the treatment recommended was foolish: *To prescribe a residence for an invalid is as unwise as to prescribe a diet; the patient's experience will be his best guide in both matters. Unfortunately, however, when a person is so far gone as to consult a physician about either, he has lost his common sense, and he too frequently persists in obeying such prescription where common sense would earnestly urge him to disobey* (F. Porcher 1943, 206–208).

Others of his class shared F. A. Porcher's opinion of the medical profession. Dr. Peter Cordes Porcher, who was studying medicine in Paris, wrote to his mother, Elizabeth Sinkler Dubose, at Ophir on April 17, 1837: *I never took any real interest in medicine until I came here; for really our manner of teaching it in the United States is so much of quacks, you see so little science in the profession, what it is* [American medicine], *is quite trash, and time lost; but here you have men who are prodigies of learning, men who have grown old and blanched in the walks of wisdom* (P. Porcher).

F. A. Porcher had an ongoing friendship and respect for his Yale classmate, Morton Alexander Waring, who had become a physician. Dr. Waring was the son of Col. Morton A. Waring, a prominent Charleston businessman and Ashley River plantation owner. His mother was Rebecca Hamilton, daughter of Paul Hamilton, Secretary of the Navy under President James Madison. Dr. Waring's siblings were Paul Hamilton Waring, who owned Buck Hall Plantation just south of Pineville; Rebecca B. Waring; Susan Mary Waring; Mary Hamilton Waring; and Col. Archibald Hamilton Waring.

Dr. Waring graduated from the Medical College of South Carolina in 1829. He purchased Oakfield Plantation around 1830 with plans to practice medicine there and reside in Pineville during the summer (F. Porcher 1943, 84). In the summer of 1831, Dr. Waring and his wife, Anne Henrietta Rhodes, were living in Pineville. Anne Waring's sister, Rebecca Branford Rhodes, was visiting. At a ball on July 4, Rebecca was introduced to F. A. Porcher, whom she later married (F. Porcher 1943, 83). So in addition to being friends, Porcher and Waring became brothers-in-law.

In his memoirs, Porcher briefly describes Waring: *He was now* [1844] *the acknowledged head of his profession in the country; his business was very extensive. His wife* [Anne, Porcher's sister-in-law] *always remembered the tie that had once connected us and was ready at all times with offers of friendship and good will towards my family.*

Dr. Waring, who later owned and planted Chelsea Plantation, was quite active in St. John's during the Civil War, tending to the medical needs of those on the plantations. As a physician during the war, he was able to move freely and played a significant role in disseminating news of the war.

Rev. Robert Wilson, MD, gave a personal description of Dr. Waring in his address to the St. John's Hunting Club on July 4, 1907. Dr. Wilson recalled meeting Dr. Waring years before at Lewisfield Plantation as he awaited a ride to Pooshee: *Once I remember being stranded, by some mistake, there being no buggy to meet me. The elder Dr. Morton Waring, however, promptly made room for me in the sulky, which seemed hardly capacious enough for his huge frame. It was characteristic of this sport-loving and obliging gentleman that seeing a covey of partridges across the road he insisted on my getting down for a shot, and scolded me roundly for riding with an unloaded gun and having no ammunition except what was in my valise.*

Later in the same address, Dr. Wilson speaks about Dr. Waring's son, Dr. Morton N. Waring, and the love of sports that father and son shared: *I recall the time when only two men of this neighborhood, the younger Thomas Porcher of Whitehall and the second Dr. Morton Waring, kept bird-dogs, though many deer hunters were good wing shots. These two also, were practically the only turkey hunters* (Wilson 1907).

At the height of his career as planter and physician, Dr. Morton A. Waring must have had a good income because he was able to send his son, Morton N., to medical school in Paris. This fact was discovered in a letter from John Safford Stoney, a medical student in Paris, to his mother at Medway Plantation on November 30, 1856: *I found Morton living over the other side of the river . . . not knowing anyone, he had not yet learned our cheap manner of living in the Latin Quarter* (Stoney 1856).

Dr. Morton A. Waring and his family said goodbye to the St. John's area after the Civil War ended. T. P. Ravenel noted in his journal of September 30, 1869: *Dr. Morton Waring's family left for Florence where he goes to practice medicine after a residence* [and] *practice of 40 years in St. John's & few physicians have enjoyed more the confidence of a community than Dr. Waring has.* No one knows why, at age sixty, Dr. Waring left the community where he was so well respected. I suspect that he did not have much in the way of savings and that the people of St. John's were too poor to continue to support his medical practice. In 1873, he sold Chelsea Plantation. He died in Florence in 1875.

Dr. Morton N. Waring returned from his medical studies in Paris in 1857 and, in 1858, married his first cousin Rebecca H. Webb of Beaufort. Morton N. Waring practiced medicine with his father in Pinopolis until 1861 when he left to serve in the Confederate Army. He returned to his Pinopolis practice in 1866. He died on September 21, 1882, at age fifty-seven.

On October 8, 1882, the furniture and other property of the estate of Dr. Morton N. Waring were sold. His widow, Rebecca H. Waring, and daughter, Eliza, left Pinopolis to reside in Summerville (T. Ravenel). Thirty-seven years later, in 1919, they were destitute and living at 36 Coming Street in Charleston. It was not until 1919 that South Carolina passed legislation to give financial aid to needy widows of veterans of the Civil War. Rebecca applied on November 3, 1919, and attested that she had no income and no property. Her application was approved, and she was listed as an Honor Roll Widow.

The village of Pinopolis settled by Waring and Porcher became a strong social summer community. Diaries describe frequent dances, parties, and picnics. Visiting among local residents and those from other villages was constant. It is documented that, by 1860, fourteen of the plantations of Middle St. John's had summer homes in Pinopolis. Journals and diaries frequently mention the presence of other homes whose locations cannot be identified.

A major social event for the men of St. John's was the monthly meeting of the St. John's Hunting Club. Although hunting was part of their initial pur-

This clubhouse for the St. John's Hunting Club was built in 1942. See pages 196 and 197 for additional St. John's Hunting Club photographs.

pose, the meetings soon became entirely social. Plantation owners of St. John's founded the St. John's Hunting Club on May 1, 1800. Initially, there were twelve members; each was to provide a meal at meetings on a rotating basis. They met under the shade of an oak tree near Pooshee Plantation until the first clubhouse near Black Oak was completed in 1801. A new building was completed in 1811, and meetings were held there until 1861.

Members present for the December 1861 meeting were Drs. Morton A. and Morton N. Waring; William H. Cain; Joseph P. Cain; W. F. Ravenel; Dr. Rene Ravenel; Thomas P. Ravenel; John Henry Porcher; Julian H. Porcher; and A. M. Porcher. There were no other meetings during the Civil War.

The clubhouse was burned in 1865 by Federal troops when General Potter and his raiders were camped at Pooshee (Orvin 1973). When the club was reorganized in 1866, meetings were held at various plantations. Meetings were suspended in 1869 when members could no longer afford to provide lavish meals.

A reorganization meeting took place in 1900 at Pooshee Plantation to celebrate the one-hundredth anniversary of the club. Six members from the 1869 roll were present. Meetings were moved to Indian Field Plantation from 1904 to 1915, to Wampee from 1915 to 1933, and then back to Pooshee until the plantation was flooded in 1941 by Santee Cooper's creation of Lake Moultrie. The only meeting ever held in Pinopolis was held at the residence of Percival R. Porcher, the Chapel Hill Plantation summer house, in October 1941. At that meeting, members noted that every plantation home of Middle St. John's had been flooded.

Percival R. Porcher III and his brother, Richard Dwight Porcher, donated land on a portion of the Pooshee property, just north of what is now Bonneau Beach, that had not been flooded, so that a new clubhouse could be built. The first meeting in the new clubhouse was held on May 23, 1942. The club has met there ever since except when meetings were suspended during World War II. The midday dinner meetings are now held every May and October. Club records show that, in July 1907, the cumulative number of members was 125, at least 45 of whom had lived in Pinopolis.

Most of the men of Pinopolis and St. John's joined the Confederate cause during the Civil War. Pinopolis made its own contribution to the Confederacy with a niter factory, which was located halfway between the village of Pinopolis and Wampee Plantation. Niter, also known as saltpeter, has the chemical name of potassium nitrate. When mixed with sulfur and charcoal, it makes gunpowder. In 1862, as the Confederacy was running out of gunpowder, Col. James Chestnut, chief of the Confederate Military Department, appointed Professor S. F. Holmes of South Carolina College to be in charge of the niter project for South Carolina. Holmes enlisted Pinopolis resident Thomas Porcher Ravenel, who was living in the Woodboo summer house, to direct the Pinopolis Niter Works: *Received a letter from Professor S. F. Holmes offering me an appointment under him in the Nitre and Mining department to undertake the making of Saltpetre in this section* (T. Ravenel).

Joseph LeConte, professor of chemistry and geology at South Carolina College, was enlisted to write a paper on how private citizens could assist in making niter. LeConte noted that it could be found naturally where there was decaying organic matter in dirt cellars, stables, manure heaps, and under homes. In another process, human urine was poured over a pile of soil, raw manure, and other material. T. P. Ravenel arranged for pots to be supplied to homes, and urine was collected periodically. A legendary story persists that when the ladies were chatting on the piazza of a Pinopolis home, one might occasionally stand and say, "Please excuse me for a moment; I have to go and shoot a Yankee."

In the twentieth century, Elizabeth "Bessie" Porcher Ravenel, daughter of T. P. Ravenel, was the Pinopolis postmaster and had to take tests to become certified in the newly organized U.S. Postal Service. As part of the requirements for certification, she wrote a paper on the Pinopolis Niter Works, using as her source the Civil War diary of her father.

Pinopolis was more fortunate than Pineville during the Civil War. Pineville never recovered as a village after the war because most of the homes were burned. As T. P. Ravenel recorded in March 1865, *there were no buildings burned in Pinopolis.*

chapter IV

Pinopolis from the Civil War to Santee Cooper, 1865–1939

Nobody knows the trouble I've seen

Negro Spiritual

THE CIVIL WAR ENDED THE ERA OF THE SOUTHERN planter. St. John's Berkeley was comprised largely of planters whose plantations had provided economic security for their families as well as their slaves' families. Times were difficult in St. John's Berkeley as anywhere else in the South. Following the Civil War, poverty and disorder destroyed hopes for a speedy financial comeback. As was seen in St. John's Berkeley, some plantation owners resolved to remain on long-held family estates and to continue with the time-honored tradition of planting. Other landowners sold their property to wealthy Northern investors but remained to be employed as caretakers. Plantations up and down the Santee and the Cooper River were sold to timber companies.

Seventy-five years had passed since the Civil War, but during my childhood, I still saw the aftereffects. The old homes in Pinopolis, which had once served as plantation summer homes, were rundown, there was no shame in wearing old clothing, and, in general, austerity was a way of life. Circumstances for most had been greatly reduced.

I visited my grandparents, William Henry Sinkler and Harriet Palmer Sinkler, at their home, Eutaw Plantation, every summer until I was six years old. It never occurred to me that they were anything but country people living in plain surroundings. Theirs was a simple lifestyle with few extras.

In my later years, I realized that Eutaw Plantation was rich in history and had been a prosperous enterprise for the

Sinkler family and their dependants from 1808 to 1861. For example, the Sinklers of Eutaw had owned some of the finest racehorses in the South. Horse racing was a sport for the wealthy. The Sinklers were more than just financially able planters; they were able to loan more than $100,000 to the Confederacy during the Civil War. Unfortunately for them, this was an investment that saw no return. By the time the Eutaw Plantation dwelling house was dismantled in 1940, the only remaining evidence of past grandeur was my grandparents' fine furniture and family silver.

The table below shows how dramatically assets changed for several St. John's planters from pre–Civil War 1860 to post–Civil War 1870.

Name	Age	Occupation	Land, 1860	Land, 1870	Personal Assets, 1860	Personal Assets, 1870
Rene Ravenel	44	farmer	$10,000	$1,000	$57,000	$500
Harlock H. Harvey	63	farmer	$3,000	$900	$25,000	$850
Charles Porcher	69	farmer	$12,000	$8,000	$73,000	$3,000
T. W. Porcher	32	farmer	$35,000	$4,000	$67,000	$3,000
Joseph Palmer	52	farmer	$22,000	$1,300	$47,000	$600
A. D. Hare	35	miller	$700	$200	$3,400	$1,000

Planters, having lost their financial capital and labor force, were unable to continue operating their plantations. To add insult to injury, the U.S. census listed Ravenel and others as "planters" in 1860 but "farmers" in 1870. While assets and land values plummeted, inflation drove up the cost of goods and supplies. For example, a cow, which could be purchased for $15 in 1861, cost $150 in 1863.

The plight of the recently freed slaves is even more difficult for us to imagine. Very few former slaves owned land or had any other assets besides the clothes on their backs. With a few exceptions, most former slaves could neither read nor write. In some cases, there was a cooperative effort between landowner and former slave to plant and harvest a cash crop. Contracts drawn up between landowners and the Freedmen's Bureau were meant to benefit both. In other scenarios, however, former slaves justified stealing from their former masters who had stolen their freedom.

In March of 1865, Henry W. Ravenel of Hampton Hill Plantation near Aiken noted that the Negroes were completely bewildered at the change in their condition after the Civil War. Ravenel wrote that some were willing to remain and work for wages. Some were delirious at the prospect of good living without any responsibilities. Others did not want their newly found freedom and preferred slavery. A group of Hampton Hill Plantation former slaves met with Ravenel on March 4, 1865, to discuss proposed work arrangements. He noted that the freedmen expected the U.S. government to provide for them without any work obligations. Even after Ravenel's negotiations, no one had come to work three days later. By December 1865, Ravenel was so discouraged at not being able to find laborers that he was ready to sell Hampton Hill Plantation and move to Aiken, South Carolina. Ravenel asked $15,000 for his plantation but received an offer of only $10,000. He was reluctant to sell. Seventeen years later, in 1882, land values had dropped so much that Hampton Hill sold for a mere $2,500.

In his private journal, Henry W. Ravenel, on February 6, 1866, described an incident regarding Thomas P. Ravenel, a planter of Woodboo Plantation, and his former slaves: *Thomas has not yet been able to get any negroes to contract for work. Every one of his negroes has left Woodboo. All have gone over to Fair Spring where they are living in dirt cellars, & Markley* [owner of Fair Spring] *is to allow them land to plant. Even those who have trades prefer planting. They have a mania for owning a small piece of land & hope to make enough by planting to purchase land* (H. Ravenel 1947, 272).

Wade H. Markley (b. 1826), owner of Fair Spring Plantation, was capitalizing upon the Negroes' desire to own land by providing them with a piece of land for farming. In return, the Negroes were to turn over stolen goods to Markley for resale.

In the author's opinion, Henry W. Ravenel, although unable to understand the Negroes' desire for freedom and property, had more concern for the welfare of the Negro than did Markley. Ravenel remarked in his journal, *I think masters . . . should remain on their plantations among their negroes* [that is, not just leave and abandon them] . . . *If we believe firmly in the Divine sanction which the Bible affords to this relation, we should not be the first to sever it, by abandoning them . . . In the sight of God, we have a sacred duty to stand by them as long as they are faithful to us . . . We must reward their fidelity to us* (H. Ravenel 1947).

So, here we see a paradox. Henry W. Ravenel believed that God sanctioned slavery, yet he did not abandon what he considered the responsibilities that went with that practice. By the end of 1865, however, Ravenel had changed his belief about slavery: *I cannot doubt, however, that the abolition of Slavery must be in accordance with the Divine will* (H. Ravenel 1947, 268).

In the years following the Civil War and Reconstruction, many of the freed slaves moved to the North, seeking more opportunities for employment and education. Those who remained in St. John's subsisted primarily as field hands, mill workers, loggers, house servants, artisans, and independent farmers.

By 1930, many of the plantations had been sold to timber companies or Northerners, who used them for hunting and recreation. The Cain family of Somerset Plantation, which once had been prosperous, was reduced to supplementing their farming income by leasing hunting rights and initiating other business ventures. William Cain, who grew up at Somerset, wrote about his frustration as a teenaged boy not being allowed to hunt on his own property.

His Aunt Caroline Cain added to Somerset's earnings by successfully creating a family business. She planted jonquil bulbs, let them multiply, and sold them to markets from Charleston to New York. She increased her profits from $2.50 the first year to $500 the following year. Other women of the Cain family were enterprising as well, selling homemade apricot leather (a sweet confection).

In 1930, Berkeley County was still largely agricultural and its residents poor. The total capital investment in Berkeley County in 1930 was $113,000. The South Carolina Public Service Authority project, commonly known as Santee Cooper, was about to change the face of Berkeley County. Forty-two miles of dams and dikes were constructed to contain the Santee River, thereby creating Lake Moultrie and Lake Marion. The Pinopolis Hydroelectric Plant was built to supply power locally and to Charleston industries during World War II. The Santee Cooper project involved several counties and was funded by a $35 million federal grant. The majority of that amount was invested in Berkeley County (Edgar 1984).

Although the Santee Cooper project pumped millions of investment dollars into Berkeley County, there was a downside for those who owned land in the path of the enormous flood site. Basically all of Middle St. John's Berkeley was to be flooded by the creation of Lake Moultrie. Those plantation families who had managed to survive the Civil War seventy years earlier were now facing a second disaster. Having lost most of their financial assets except their land as a result of the war, they now faced the loss of ancestral land and family homes.

Santee Cooper purchased 1,100 individual tracts of land, containing a total of 177,000 acres, resulting in the displacement of 901 families. The lakes redefined the geographic boundaries of Pinopolis and all of St. John's. However, the project also brought prosperity. During the height of the construction, 12,500 workers, previously on the relief rolls, were employed. Santee Cooper later created permanent jobs while the lakes provided opportunities for the development of recreation, tourism, and ancillary businesses. On November 12, 1941, the reservoir began to fill, and on February 17, 1942, the Pinopolis Hydroelectric Plant produced its first electricity.

When telephones were introduced to the area in 1904, two of the first six telephones in Berkeley County went to the Pinopolis residences of J. Russell Williams and Dr. W. Kershaw Fishburne. Mary Winter Briscoe, who was involved with the initial private telephone company, acquired it in 1939 and named it Home Telephone Company. In 1947, the company passed to her daughter, Thelma, and son-in-law, Shelly Helmly. Their sons, Dozier and Bobby Helmly, are still involved in the business, which continues to thrive 102 years from its inception.

The oldest surviving telephone directory of the Home Telephone Company, from 1947, lists ninety residential customers. Seventeen of these lived in Pinopolis: Moultrie Ball; James Bright; F. E. Carson; Homer Dantzler; Rembert Dennis; Dr. Fishburne; H. H. Harvey; H. W. Harvey; Dr. W. H. Lacey; J. H.

Massalon; Louise Parker; P. R. Porcher; Herbert Stokes; J. G. Stevens; Harriett Vardell, Dr. J. N. Walsh; and J. Russell Williams Sr.

Because Pinopolis has never been incorporated, residents have been responsible for their own water and sewerage needs. Water was drawn from wells with buckets initially and later by hand pumps. It was not until 1941 that electric pumps could be used.

A. Davis Hare returned to Pinopolis, his birthplace, in 1946 after serving in the U.S. Coast Guard. He worked as a jack-of-all-trades before getting into the water works business by happenstance. In 1949, his neighbor, Harriett Vardell, had a pump problem that Hare couldn't solve. He got water to her by running a pipe across the yard from his own water supply. Soon, Edgar Davis in the Episcopal Rectory and others sought his service, so pipes were snaked throughout the village. As the demand grew, a more sophisticated system was installed. Today, the A. D. Hare Waterworks Company has three hundred clients in Pinopolis and another three hundred along Whitesville Road and Highway 16. At present, eight wells provide the needed water. About ten homes in Pinopolis still use private wells.

The Pinopolis Hydroelectric Generating Plant, which began operation on February 17, 1942. To the left is Lake Moultrie. To the right is the Tail Race Canal. *Courtesy of Jim Huff, Santee Cooper, 2005.*

chapter V

Plantations, Homes, and Other Buildings

There is something in the pineland of our Low Country which has nothing like it in any other part of the world.

F. A. Porcher

In chapters 5 through 8, map notations in text refer to the map on front and rear inside covers.

PINOPOLIS IS THE ONLY PINELAND VILLAGE IN Berkeley County with a significant number of historic homes still standing. Of the thirty-four existing historic structures in Pinopolis, only twelve were present during the prosperous period before the Civil War: the Episcopal rectory, the Chelsea summer house, Chapel Hill summer house, Wampee summer house, Wampee Plantation house, Somerset summer house, Somerset summer house lodge, the Kate Harvey house, Dr. Morton Waring's office, the Northampton summer house, the Adam Davis Hare house, and the Old Post Office, which used to be the Woodboo summer house lodge. These surviving homes have been modernized while still maintaining, for the most part, their original exterior facades. Some homes required major structural changes because of hurricane damage. Hurricane Hugo in 1989 destroyed almost every majestic pine in Pinopolis, leaving today's visitor wondering what a pineland village really looked like. Most of the wooded spaces of my childhood are filled with modern homes. There are still no commercial stores or government buildings.

During my childhood in the 1940s, most homes were rundown by today's standards, and some still had exterior kitchens. The older photographs will show how they used to look. The heart pine or cypress siding held up well without surface treatment; however, outside walls were sometimes whitewashed with a mixture of lime and water.

William Ball, MD, of Aurora, Illinois, wrote to Anne S. Ball on May 22, 1983, regarding his childhood memory of the Somerset summer house: *Thinking of Cassie* [Caroline Cain] *brings up recollections of the Cain house* [Somerset summer house], *now in such terrible condition . . . the kitchen was a log cabin with one room for cooking, and one for storage, about 10 yards behind the house, and connected to it by a platform or piles, like a bridge . . . I used to play like it was over water, making it exciting to jump off* (W. Ball 1983).

This chapter will describe existing and no longer existing plantations and their associated summer homes in Pinopolis. Also described will be historic houses, churches, and other structures not associated with plantations. Useful information about homes and buildings was obtained from maps drawn by three Pinopolis residents from the early 1900s (see page 164 for maps): Maria Jervey Ravenel; her brother, Dr. James Jervey Ravenel; and Emma Jenkins Gribbin. Soon after the settlement of Pinopolis, it was customary for a plantation owner to call his summer home "the lot."

Chelsea Plantation — no longer existing

Chelsea Plantation was well known to F. A. Porcher: *North of Northampton on the same side of the swamp is Chelsea, an old St. Julien seat if I mistake not, but like all the other seats of that family it had passed by marriage into the hands of a Ravenel. My uncle Peter Porcher who had married a lady of that house was for a time the owner, but after his wife's death it reverted to her family, and it again passed by marriage, into the hands of a Mr. Broughton. He died before I was born, but his sons Daniel and Alexander I knew very well, the former continued to be the proprietor of Chelsea until he died, when it passed by purchase into the possession of Dr.* [Morton Alexander] *Waring . . . Chelsea was one of those plantations on which the cultivation of rice was long continued; a part of the tract was known as Hepworth, on which were standing the ruins of an old brick house, said to have been the residence of one of the St. Juliens* (F. Porcher 1943, 72–73).

The construction date of the Chelsea Plantation house is uncertain. In about 1870, F. A. Porcher noted that the house was one of the oldest in the St. John's parish, being about 120 years old. That would put its construction at about 1750 (F. Porcher 1921, 40). However, writing in 1917, Mrs. Mary Rhodes Henagan, daughter of Dr. Morton A. Waring and niece of F. A. Porcher, said, *Chelsea, our plantation and winter home . . .* [was] *a big old colonial house built in 1714* (Jervey 1921).

Dr. Morton A. Waring married Anne Henrietta Rhodes on December 22, 1830. In 1836, he purchased Chelsea Plantation and made it their home (F. Porcher 1921).

On January 7, 1873, an entry in the diary of T. P. Ravenel reads: *Chelsea Plantation was sold at auction for $1520 under the mortgage of Dr. M. Waring, bought by Walkins of Charleston* (T. Ravenel). J. H. Porcher recorded on June 14, 1900, that Chelsea was sold to Mr. Purdy (J. Porcher).

Chelsea Plantation Summer House
912 Morton Waring Lane (map: 20)

Previous oral and written accounts of Pinopolis history refer to the existing structure on Morton Waring Lane as the Buck Hall summer house, the first home in Pinopolis, built by Col. Morton Waring, MD, of Buck Hall. Current research reveals that each description in the previous statement is incorrect. Dr. Morton Alexander Waring, who at the time was living at Oakfield Plantation, which he had purchased in 1831, built the first home in Pinopolis in 1834. In 1844, when he built his second Pinopolis home, the present structure, Dr. Morton A. Waring was living at Chelsea Plantation. Originally, this existing home was called "the Chelsea lot." Miss Susan R. Jervey recorded in her diary on March 8, 1865, that Rev. P. F. Stevens took a piano to the Chelsea lot, home of Dr. Morton A. Waring, for safekeeping (Jervey 1921).

Chelsea Plantation summer house as it appeared in 1970. *Courtesy of the Elias Bull Collection, Charleston County Library.*

Inset: Rear of the Chelsea summer house as it appeared in 1970. *Courtesy of the Elias Bull Collection, Charleston County Library.*

There has been confusion regarding the identities of Col. Morton Alexander Waring (1784–1863); his son, Dr. Morton Alexander Waring Jr. (1809–1875); and the doctor's son, Dr. Morton Nathaniel Waring (1831–1882). Further confusion arose because there were five women in the family named Rebecca.

Col. Morton A. Waring was a prominent businessman in Charleston who also had a plantation on the Ashley River. He died at Buck Hall Plantation near Pineville. Buck Hall was the plantation home of his son Paul Hamilton Waring. It was assumed that Col. Morton A. Waring owned Buck Hall because his tombstone at Biggin Church states that he died at Buck Hall. This is confirmed in the diary of Thomas Porcher Ravenel: *Died at Buck Hall the residence of his son P. Hamilton Waring, Col. Morton* [Alexander] *Waring of St. Paul's in the 80th year of his age* (T. Ravenel).

Col. Morton A. Waring was not a physician but was mistakenly listed as such by Dr. J. I. Waring (Waring 1923). This error led Pinopolis historian Margaret Stevens to conclude that it was Col. Morton Waring who built the first home in Pinopolis when, in fact, it was built by his son, Morton Alexander Waring, MD. Adding to the confusion, Morton Nathaniel Waring was also a physician and lived in Pinopolis in the Chelsea summer house.

It was common in diaries and journals of this period to use only first names. This makes it difficult in research to identify individuals, and it is likely that Mrs. Stevens's confusion was amplified by the names of the women in the Waring family.

For example, there were five Rebeccas:

- Col. Morton A. Waring married Rebecca Hamilton (1786–1871).
- They had a son, Morton A. Waring Jr., and a daughter, Rebecca B. Waring (b. 1808).

- Dr. Morton A. Waring Jr. married Anne Rhodes whose sister, Rebecca, married F. A. Porcher.
- Morton A. Waring Jr. had a daughter, Rebecca Hamilton Waring (b. 1834).
- Morton A. Waring Jr. had a son, Morton N. Waring, who married Rebecca Hamilton Webb (1835–1920).

Often quoted documents state that Dr. Morton Waring built the first summer home in Pinopolis in 1834. Until now, the present Chelsea summer house, previously called the Buck Hall summer house, was thought to be that original house. On the contrary, the present Chelsea summer house is actually Dr. Waring's second home. Richard Porcher wrote, *in 1834 the first two log houses* [in Pinopolis] *were erected . . . by Dr. Morton Waring and Mr. F. A. Porcher.* Since the present house is built of planks, it clearly could not be the log house referred to by F. A. Porcher in his memoirs. Porcher also recorded in 1844: *Of the two houses which had been built by Dr. Waring and myself ten years before, his only was standing, and with considerable enlargements it was now the residence of Henry W. Ravenel* [1814–1870]. The location of Waring's first home, built of logs and still standing in 1844, is unknown. The second Pinopolis home built by Dr. Morton A. Waring on the Chelsea lot is the existing structure on Morton Waring Lane.

When F. A. Porcher said, *Dr. Morton Waring and his family lived in a house which he had built at the north of mine*, he was referring to the second home built by Dr. Morton A. Waring. Likewise, Porcher is referring to his second home when he used the word "mine" (F. Porcher 1943, 102).

Inset: Living room fireplace of Chelsea Plantation summer house.

Chelsea Plantation summer house in 2005.

In 1869, Dr. Morton A. Waring moved to Florence, South Carolina. His presence there is documented in the 1870 U.S. Census for Darlington District.

The Chelsea Plantation summer house lodge dining room in 2005.

On December 23, 1872, Dr. Morton A. Waring of Florence passed the Pinopolis summer home to his son, Morton N. Waring, MD. Morton N. Waring studied medicine in Paris in 1856 (Stoney 1856). He returned to Pinopolis in 1857 and began the practice of medicine. Ravenel records that Dr. Morton N. Waring married Rebecca Hamilton Webb in 1857 and that they were living in Pinopolis in 1858 (T. Ravenel). The 1860 St. John's census confirms that Dr. Morton N. Waring, age twenty-eight, was a resident of St. John's. During the Civil War, Dr. Morton N. Waring was a surgeon and a private in the Army's Fourth Cavalry Company K, Charleston Light Dragoons.

On November 19, 1878, Morton N. Waring deeded the Pinopolis property of one and one-half acres of land to his wife for $200. She is listed in some genealogy records as Rebecca Ladson Webb Waring, but her documented name is Rebecca H. Waring. Dr. Morton N. Waring died in Pinopolis in 1882 at age fifty-seven.

On February 2, 1883, Rebecca Hamilton Waring deeded the property to William Francis Ravenel (1828–1906) for $425. In 1901, William F. Ravenel deeded the property to his daughter, Miss Annie Stevens Ravenel (1865–1954). Miss Annie Ravenel's health was fragile, so in 1915, she sold the property to Josephine H. Pegues and moved to Tryon, North Carolina.

Josephine Pegues enlarged the property through the purchase of the Tennis Court Lot (map:Ⓝ) at the corner of Pinopolis Road and Morton Waring Lane. Old deeds reveal that it was called the Tennis Court Lot when conveyed to Mrs. Pegues by Minnie Harvey (Mrs. Lewis H. Harvey). The Tennis Court Lot was later sold to Mrs. Lucas Padgett, who built the home now owned and occupied by Steve and Mary Nettles.

Josephine Pegues sold the Chelsea summer house to Janie Caddell. In 1924, Janie Caddell sold it to Mrs. Joseph E. Edwards, who left it to her daughter, Margaret E. Stevens, who initiated the search for the builder of the Chelsea summer house.

A previous owner of this house, Miss Annie Ravenel, returned to Pinopolis in 1945 and rented three rooms of the Chelsea summer house from Mrs. Edwards. Miss Annie lived there for nine years until her death in 1954. I remember Miss Annie quite well because she used to give cookies to me and other children from her back steps. This cookie ritual with the village children is well described by Anne S. Fishburne in her book *William Kershaw Fishburne MD* (Fishburne 1969).

The present owners of the Chelsea summer house are Eugene and Sherry Villeponteaux.

Chelsea Summer House Lodge

1918 Pinopolis Road (map: 21)

Some summer homes, like Chelsea and Somerset, had a lodge, which served as a dormitory for boys in the family. Often, male visitors also stayed in the lodge.

In 1924, Mrs. Joseph E. Edwards gave the Chelsea summer house lodge to her daughter Margaret and husband, Goodlow Stevens, as a wedding gift. They moved the Chelsea summer house lodge a few yards to the north and added two cabins, which were already on the Chelsea property, to the lodge to create the present home (interview by author with Robert Stevens).

Present owners Gerald and Nancy Stuckey have made extensive improvements. The two rooms of the old lodge are now the entry hall and dining room.

Dr. Morton Waring Office

901 Morton Waring Lane (map: 14)

According to Margaret Stevens, Dr. Waring's office was originally on the site of Waring's home on the Chelsea lot. Dr. Richard Y. Dwight moved the office to his lot, which was formerly the second Pinopolis home of F. A. Porcher. Dr. Dwight placed the office on poles about six feet above ground. Henry R. Dwight Jr., the

Dr. Morton Waring used this medical office soon after he settled in Pinopolis in 1834.

doctor's grandson, remembers that his grandfather's patients could sit in the shade under the office while waiting to be seen.

Around 1918, Dr. Dwight's granddaughter, Rowena Ravenel Dwight (b. 1909), known as Nina, used a part of the office to give school lessons to her black friend, Mary Brown (b. 1917), also known as Dolly. In a later chapter of this book, Dolly Brown tells the story of her close association with the Dwight family.

Robert LaFaye, who owned the lot just north of Dr. Dwight's lot, purchased the Dr. Morton Waring office from the great-grandson of Dr. Dwight, Edgar Davis Jr. The building, which was quite near the boundary line between the Dwight and LaFaye lots, was simply moved across the line, rotated 180 degrees, and lowered to ground level.

Elliott Williams Bishop, the present owner, converted the office into small living quarters.

Frederick Adolphus Porcher House — no longer existing

1860 Pinopolis Road (map: Ⓚ)

The second Pinopolis home of Frederick A. Porcher, built in 1844, is seen as it appeared in 1939. *Courtesy of the South Caroliniana Library.*

Frederick A. Porcher settled Pinopolis with Dr. Morton A. Waring in 1834. Since Porcher's family already had a summer house in Pineville, he assisted Waring in constructing his home before building his own in Pinopolis. He probably spent little time in his first Pinopolis home. As he wrote in 1844, *I began to make arrangements for building a house* [second house] *in Pinopolis. At first I contemplated only a refitting of the old house, but on examination it was found impracticable and a new house was resolved upon* (F. Porcher 1943, 101). The location of Porcher's first house is unknown. When he built his second summer home in 1844, he was established at Somerton Plantation.

Dr. Richard Yeadon Dwight purchased Porcher's house, probably in the 1860s. Dwight is listed as a physician, age seventy-two, in the 1910 U.S. Federal Census. Henry Ravenel Dwight (1873–1961), author of *Some Historic Spots in Berkeley*, lived in the Frederick A. Porcher house his entire life. He was the son of Dr. Dwight and Rowena Ravenel (1837–1924).

Dr. Richard Y. Dwight's grandson, William Edgar Davis Jr., acquired the property that was originally the second home of F. A. Porcher. In 1950, Davis determined that it was not reasonable to preserve Porcher's second Pinopolis home and built a new home on the same spot.

Cedar Spring Plantation — no longer existing

F. M. Kirk wrote in his "Cedar Spring History," *The plantation* [Cedar Spring], *located eight miles from Pinopolis, was formerly known as Buffalo Hole, from the curious*

Cedar Spring Plantation house, the birthplace of Frederick A. Porcher, as it appeared in 1939. *Courtesy of the South Carolina Historical Society.*

sinks to be found there and elsewhere in St. John's. Philip Porcher (1730–1800) of Old Field Plantation in St. Stephens bought Cedar Spring Plantation shortly before the Revolutionary War and gave it to his youngest son, George Porcher (1775–1813).

Frederick A. Porcher (1809–1888), son of George Porcher, wrote in his memoirs (1943) that his father in some ways felt cheated that plantation land in St. John's had been made his inheritance while all the other Porcher sons had received valuable land along the Santee River. After a series of freshets claimed the fertile growing fields of the Santee River plantations, George Porcher was glad that he had established himself in St. John's.

M. C. Orvin mentions Cedar Spring Plantation and its owner George Porcher as early as 1800 in *Historic Berkeley County, South Carolina, 1671–1900* (1973). Orvin says, *On May 1, 1800, the St. John's Hunting Club was organized . . . Mr. George Porcher of Cedar Spring Plantation arranged the first dinner.* George Porcher built the plantation dwelling house at Cedar Spring in 1804. He married Marianne Gendron (1781–1835), widow of Gabriel Gignilliat and daughter of Capt. John Palmer, in 1806. They had four children.

Their two sons, Frederick A. Porcher and Dr. John Palmer Porcher (1812–1839), played key roles in Cedar Spring Plantation's history, as well as St. John's history. Frederick A. Porcher in his memoirs remembers a close relationship with his father and remembers also the pleasures of growing up at Cedar Spring Plantation. George Porcher died in 1813. His widow enlarged the Cedar Spring Plantation dwelling house in 1825.

The Porchers' first son Frederick, who has already been much quoted through his memoirs written in 1867, established himself as a planter at Somerton Plantation before leaving that occupation and becoming a college professor and a man of letters. Second son John Palmer inherited Cedar Spring Plantation but chose to study medicine and later became a physician. Dr. John P. Porcher married Anna Ford Banks in 1835 and died in 1839, leaving a widow and two young boys, George and John P. Porcher Jr., to be cared for by family members.

Cedar Spring Plantation was left to Dr. Porcher's sons, to be held in trust by their uncle, Frederick A. Porcher. The doctor's widow, Anna Porcher, died in 1844 and named Frederick A. Porcher as her sons' legal guardian. Because of their young age, Frederick Porcher largely left the two boys in the care of their maternal grandmother in Charleston. Porcher intended to assume the responsibility of managing Cedar Spring Plantation himself until his nephews were of age but instead passed the plantation's care over to his first cousin, Henry Francis Porcher (1819–1860).

John Palmer Porcher Jr. (1836–1864) returned to St. John's from Charleston to begin his planting career at Cedar Spring Plantation in the mid-1850s. He married Catherine "Kate" Cordes Deveaux (1839–1923), daughter of Steven Gough Deveaux, and with Catherine had two sons. Both sons died at an early age. John Palmer Porcher Jr. died in 1864 at age twenty-eight. His widow, Kate, inherited Cedar Spring Plantation and the Pinopolis summer house.

John Henry Porcher recorded in his diary for January 27, 1904: *My nephew Isaac* [de Cherigny Porcher] *bought Cedar Spring Plantation from Mrs. Porcher* [Mrs. Kate Porcher] *widow of John Porcher for $10,000. He inherited from his father. It was bought by my grandfather Philip of the Old Field previous to 1800.*

F. M. Kirk confirms this transaction by explaining that Mrs. Kate Porcher, widow of John Palmer Porcher Jr., sold Cedar Spring Plantation to Isaac de Cherigny Porcher (1859–1933), grandson of Isaac Porcher III. Isaac Porcher III, in turn, was a brother of George Porcher, the original owner of Cedar Spring Plantation. Isaac de Cherigny Porcher's nephews, Percival Porcher III (1897–1965) and Richard Dwight Porcher (1900–1973), were the last owners of Cedar Spring Plantation before the property was flooded by the Santee Cooper Project in 1939.

Cedar Spring Plantation Summer House — no longer existing
(map: BB)

In a letter to his brother, Peter C. Porcher, who was studying medicine in Paris in 1837, Thomas Porcher of White Hall Plantation wrote news of St. John's people and plantations. Of particular interest is Thomas Porcher's account of summer homes being built in pineland settlements, such as Pinopolis, Pineville, and Jack's Hole: *In consequence of having so much to do I have rented out our house in Charleston, and have removed the House which young Mr. Marion built on a part of "Goshen" to the Pineland of Doctor Waring and Mr. Cain* [Pinopolis], *where I expect to live until July and then go to the back country.*

Pineville will almost be entirely deserted this Summer. James Gaillard is to go there until we go away, and then he is to take my house in Pinopolis for the remainder of the Summer.

Charles DuBose is also building there [Pinopolis], *and F. A. Porcher and Dr. John Porcher are also to be there until July and later. S. Porcher Gaillard, Mr. Isaac Porcher and Charles Snowden are building on a part of "Pooshee" near the old Butler Road, and not far from where there was a development a few years ago at a place known as "Jacks Hole."*

Dr. John Porcher and his brother, Frederick, are mentioned in Thomas Porcher's letter as going to "the Pineland of Doctor Waring and Mr. Cain." Frederick Porcher's pineland summer house was directly across the road from the Cedar Spring summer house of his brother John. In 1855, a portion of the Cedar Spring lot was given to the Episcopal Church in Pinopolis for a rectory (parsonage).

Both John and Frederick Porcher were spending their summers in Pinopolis by 1836. John Porcher died in 1839, and his wife, Anna, died five years later in 1844. Their two sons, George and John P. Porcher Jr., spent the majority of their youth in Charleston with their grandmother, Mrs. Banks. There are no records indicating that the Cedar Spring Plantation summer house in Pinopolis was occupied by the sons until 1859 when Thomas Porcher Ravenel mentions both sons in his personal journal: *November 13, 1859. Died in Charleston at the residence of his Uncle John Bryan, "George Porcher" in his 21st year of age, second son of the late Dr. John Porcher of Cedar Spring, St. John's Berkeley.*

Thomas Porcher Ravenel recorded various social activities taking place in Pinopolis in 1859:

July 1st Dancing party at Mr. Cain's [Somerset summer house].

July 5th Dancing party at Father's [Dr. Henry Ravenel's Woodboo summer house].

July 7th Dancing party at John P. Porcher's [Cedar Spring summer house]—*Gaiety & dancing, wagon driving, & etc. seem to have taken possession of our village.*

In 1859, John P. Porcher Jr. and wife Catherine "Kate" Cordes Deveaux were living at Cedar Spring Plantation and at the Pinopolis summer house. Mrs. Kate Porcher and her young son, George, spent the Civil War years at Cedar Spring Plantation. The diaries of Miss Charlotte St. Julian Ravenel and Miss Susan R. Jervey of Middle St. John's record the horrors endured by Mrs. Porcher and George at Cedar Spring Plantation during General Potter's raid in March of 1865 (Jervey 1921). Thomas Porcher Ravenel in his journal gave a last entry about the John P. Porcher Jr. family: *Sept. 2, 1869 Died in Pinopolis, George Porcher, son of the late John P. Porcher & Kate Deveaux.*

Mrs. Kate Porcher remained in Pinopolis after the Civil War and the death of her young son. Her Cedar Spring Plantation summer house was mentioned again in Thomas Porcher Ravenel's journal in 1886 as being burned in a great fire in Pinopolis. Afterwards, she moved to Somerset Plantation to live with her friend, Anna Maria Cain, and her family. "The Treasure House," which was part of the Cedar Spring Plantation summer house lot, was probably used by Mrs. Porcher to store any possessions saved from the fire of 1886.

Chapel Hill Plantation — no longer existing

According to F. A. Porcher, Chapel Hill Plantation derived its name from a chapel of ease that existed on the plantation site about 1770. William Brickell had a home there at the beginning of the nineteenth century. Isaac Porcher III

(1778–1849) purchased Chapel Hill in 1818 and lived there until his death in 1849. Isaac Porcher III had previously tried planting at Dover Plantation, which had been given to him by his father, Philip Porcher of the Old Field Plantation. Isaac Porcher was unsuccessful and blamed his unprofitable planting venture at Dover Plantation on inferior soil. He purchased Chapel Hill Plantation in hopes of a more successful yield.

Isaac Porcher III was married three times and had children by all three wives. Porcher and his first wife, Mary Weston (1781–1808), had a son and named him Isaac, making him the fourth Isaac Porcher (1803–1809). After Mary Weston Porcher's death in 1808, Isaac Porcher III married Mary Stevens (1793–1824). They had a son and named him Isaac Porcher V (1812–1849). After Mary Stevens Porcher's death in 1824, Isaac Porcher III married Charlotte Mazyck Ravenel (1795–1835), his third and last marriage. Isaac and Charlotte Porcher had two sons. Their first-born son, Percival Ravenel Porcher (1829–1864), died near Richmond, Virginia, during the Civil War. Their second son, John Henry Porcher (1835–1910), enlisted in the Confederate Army but was unable to continue because of ill health and so sought civil service.

Percival Ravenel Porcher married Maria Catherine Ravenel (1832–1907) in 1856 at Pooshee Plantation, her father's home. Percival and Maria had two sons, Percival Ravenel Porcher II (1861–1931) and Isaac de Cherigny Porcher (1859–1933).

In 1864, Maria Catherine Ravenel Porcher, sister of Thomas P. Ravenel, was living at Moorefield with her two small children when her husband, Percival Ravenel Porcher of the Charleston Light Dragoons, died near Richmond, Virginia, as the result of wounds received during the Civil War. F. M. Kirk relates: *All of the men* [in the family] *were in service, so Mrs. Porcher sent P.R.'s* [Percival Ravenel's] *body servant, Robert, to bring the body back. While in Richmond, Robert could find no way to get the body back. He is said to have wandered the streets of Richmond shouting "My young master, Mr. Percival Porcher, of South Carolina, lies dead in a hospital* [four miles away], *and I want to help to carry him back to his young wife* [Maria]." *An officer passing, hearing the name Porcher, and recognizing it as a relative, took him to General Lee's headquarters. There, an order was procured for artillery horses to be sent for the body and a guard of honor to escort the body. Due to the faithfulness of his slave, Percival Porcher lies buried in the family plot at Black Oak.*

Graves originally located at Black Oak Cemetery were relocated to Pooshee Cemetery near Bonneau before Santee Cooper flooded the area. Percival Porcher's faithful body servant Robert, later known as "Old Bob," has an unknown grave, but his tombstone was for a long time on the grounds of the Porcher beach home on Lake Moultrie. "Old Bob's" tombstone was moved to the grounds of the St. John's Hunting Club at Pooshee. Porcher family members have agreed to move "Old Bob's" tombstone to Pooshee Cemetery where he can rest beside his former master, Percival R. Porcher.

The Chapel Hill Plantation house was destroyed by fire in 1837 and rebuilt by Isaac Porcher III on the same spot. In 1849, after Isaac's death, Chapel Hill Plantation was sold to Samuel Foxworth.

Chapel Hill Plantation Summer House

931 Morton Waring Lane (map: 22)

Anne S. Ball calls this Pinopolis house the Indian Field Plantation summer house and says it was built around 1836 by Isaac Porcher III (1778–1849), but other historians have identified the summer house as being built in 1844 by Isaac Porcher III of Chapel Hill Plantation. Elias Bull (Inventory 1979) and Richard D. Porcher state that the summer house in question was built in 1844. In handwritten notes about summer houses in Pinopolis, Elias Bull calls this house the Chapel Hill Plantation summer house. In his "A History of Pinopolis," Richard Porcher called this same summer house the Chapel Hill lot.

F. A. Porcher records: "*My Uncle Isaac Porcher* [III] *was now* [1844] *a resident of the village* [Pinopolis]. *He was somewhat lame from a fall that he had had some years before, and being constitutionally indolent, he had given up the management of his business entirely to his sons Isaac and Philip. All lived together . . . My uncle was eminently social, and dearly loved a chat with a neighbor. I very often visited him, and he visited me, and both were the better for this interchange of visits* (F. Porcher 1943, 102–103).

At the death of Isaac Porcher III in 1849, Thomas Porcher Ravenel bought the Chapel Hill Plantation summer house. Ravenel lived in the summer house at least until 1858 (T. Ravenel). It is unknown exactly how the Chapel Hill Plantation summer house reverted back into Porcher hands, but Isaac de Cherigny Porcher (1859–1933), Isaac Porcher III's grandson, is shown as owning the Chapel Hill Plantation summer house after Thomas Porcher Ravenel.

Isaac de Cherigny Porcher willed the summer house to his niece Marguerite Hamilton Ravenel (1898–1965), wife of his nephew Percival Ravenel Porcher

Isaac Porcher built the Chapel Hill Plantation summer house in 1844, seen here as it appeared in 1970. *Courtesy of the Elias Bull Collection, Charleston County Library.*

III (1897–1965). Marguerite and Percival raised their three children there and lived there until 1965 when Marguerite and Percival were killed in an automobile accident. Their children, P. R. Porcher IV (b. 1931), Margaret Deas Porcher (b. 1933), and Heyward Ravenel Porcher (b. 1940), were contemporaries of mine in Pinopolis and have all given valuable help with this book.

Various families rented the Chapel Hill Plantation summer house from 1965 until 1978 when Eugene and Lynette Oliver, the present owners, purchased it. The house was significantly damaged during Hurricane Hugo in 1989. During post-Hugo repairs, the Olivers did extensive remodeling, making the old summer house into a comfortable year-round home.

The Chapel Hill Plantation summer house, seen in 2005, shows changes after repairs from Hurricane Hugo in 1989. Dormers have been added and the piazza lengthened.

Inset: Piazza of the Chapel Hill summer house.

Northampton Plantation — no longer existing

The first owner of Northampton Plantation is thought to be Pierre St. Julien, a French Huguenot immigrant. The house was built around 1716 by Benjamin St. Julien, grandson of Pierre, who had no children. It came into ownership of Gen. William Moultrie through Pierre's son, Peter, who willed it to his sister, Elizabeth St. Julien, wife of General Moultrie.

The Santee Canal was built along the northeastern boundary of Northampton Plantation at its border with Pooshee Plantation. When the Santee Canal opened in 1800, General Moultrie held a breakfast at Northampton in honor of Major Senf, the chief engineer. General Moultrie's son, who inherited Northampton, died at an early age unmarried. The property was sold many times (Kirk).

The Northampton house burned in 1844 and was rebuilt in 1846 (J. Porcher). Henry William Ravenel (1814–1887), the noted botanist, son of Dr. Henry Ravenel (1790–1867), purchased Northampton from Theodore Dubose. In February 1853, he sold the property to Henry LeNoble Stevens (1827–1862) for $10,000 and moved to Aiken (H. Ravenel 1947, xvi). During the Civil War, Henry L. Stevens served as aide to Col. P. F. Stevens (see Trinity Church) and was killed in the second battle of Manassas. Henry L. Stevens's daughter, Henrietta LeNoble Stevens, was born three months after his death. Henry L. Stevens's widow, Henrietta S. Gaillard (b. 1829), later left Northampton to their son, Charles (1854–1928), who married Mary Wharton Sinkler (b. 1857) of Belvidere. Charles and Mary Stevens lived at Northampton in the winter and in the Pinopolis home in the summer.

From 1867 to 1871, Maria Ravenel Porcher, widow of Percival Porcher, lived at Northampton with her brother-in-law, John Henry Porcher, who was renting and planting there (T. Ravenel).

Available published records do not show that Percival Ravenel Porcher II (1861–1931) owned Northampton. However, John Henry Porcher recorded in his journal in January 1904: *my nephew Percival Porcher has bought Northampton from Charles Stevens for $7,000.* Percival Porcher was living at Indian Field at the time and moved to Northampton the next month (J. Porcher).

After 1929, Percival R. Porcher II sold Northampton to A. M. Barnes and Clarence Dillon, who also bought Indian Field and used both plantations for hunting. Percival R. Porcher III (1897–1965) and his family lived in the overseer's house at Northampton and managed both properties for Barnes and Dillon. Percival IV and his sister, Margaret "Punky," remember living at Northampton in the winter and at the Chapel Hill summer house in Pinopolis during the summer.

The first Northampton Plantation home was built in 1716 and was once the home of General William Moultrie. It burned in 1844, and the building shown here was built in 1846. *Courtesy of the South Caroliniana Library.*

Their mother, Marguerite Hamilton Ravenel (1898–1965), inherited the Chapel Hill summer house from her uncle, Isaac de Cherigny Porcher (1859–1933).

Charles Stevens built the Northampton Plantation summer house, probably in the 1880s. *Courtesy of Virginia Seay.*

Northampton Plantation Summer House
2008 Pinopolis Road (map: 30)

The construction date of the Northampton summer house is unknown. Charles Stevens (1854–1928) and his wife, Mary Sinkler of Belvidere, who spent their summers in Pinopolis, likely built it. Harold H. Harvey, who grew up in Pinopolis in the 1920s and 1930s, remembers that, during this time, the Northampton summer house was vacant and the windows shattered by slingshots. During the construction phase of Santee Cooper, however, Mrs. Goodwin ran a boarding house in the Northampton Plantation summer house for workers. Virginia and Pitman Seay purchased and remodeled the house in the 1950s.

The present owner is R. W. Thornley.

Pooshee Plantation — no longer existing

Pierre St. Julien, who also owned Northampton, received a 1,000-acre land grant for Pooshee in 1705. He sold it to his brother-in-law, Henry LeNoble, who deeded it to his son-in-law, Rene Louis Ravenel (b. 1694), son of the immigrant, Rene Ravenel (b. 1656).

The first plantation dwelling at Pooshee was built in 1716 and the second in 1804. Dr. Henry Ravenel (1790–1867), father of Henry William Ravenel (1814–1887), the botanist, owned Pooshee. Dr. Ravenel added a western wing to the house in 1852. His landholdings at one time extended for fourteen continuous miles.

Dr. Henry Ravenel was one of the most efficient planters in St. John's. Henry Edmund Ravenel (1824–1863) states, *The* Southern Agriculturist *for July 1831 has a detailed account of the management of Pooshee, written by the editor, Mr. J. D. Legare. He regarded the various operations there carried on as a model for planters. He showed that under Dr. Ravenel's system, the productiveness of his lands had been doubled in the course of eleven years . . . A peculiarity of his system was that he did not rotate his crops on Pooshee but increased the fertility by heavy application of manure, produced at home, so that in the year 1831 there were hauled out and spread upon the fields of his plantation an amount of stable manure equal to 4,448 single horse carts* (Ravenel Records 1898).

Dr. Rene Ravenel (1826–1875), son of Dr. Henry Ravenel and his second wife, Elizabeth Catherine Porcher (1798–1846), built the Pooshee summer house in Pinopolis in 1856. Thomas Porcher Ravenel (1824–1898), whose journal is quoted throughout this book, was Rene's brother, and each was a half-brother of Henry William Ravenel, the botanist.

In October 1906, Percival Porcher II and his brother Isaac purchased Pooshee for $25,000 (J. Porcher). At the time of the Santee Cooper flooding in 1941, Pooshee consisted of about 4,000 acres and was owned by the sons of Percival Porcher II, Richard Dwight Porcher and Percival Ravenel Porcher III. The plantation house had been unoccupied for many years (Kirk).

Pooshee Plantation Summer House

919 Dooley Lane (map: 27)

Dr. Rene Ravenel and his wife, Isabelle M. Deveaux, built the Pooshee summer house. Rene's brother, Thomas Porcher Ravenel, recorded in his journal: *June 5–6, 1856, Rene's house in Pinopolis is raised* (T. Ravenel). When Dr. Ravenel died, Isabelle married Charles Macbeth. While it was common knowledge that this home later burned, no one knew when it burned or when the second house (the present house) was built.

Dr. Ravenel's granddaughter, Maria Jervey Ravenel (1892–1991), lived in the adjacent home with her parents, Rene Ravenel (1856–1901) and Maria Stevens Jervey (1856–1929). Late in her life, Maria, who was called Marie, drew a map of Pinopolis as she remembered it in 1900. Her map has been most useful in the research for this book. Marie said that her grandmother (Isabelle Ravenel Macbeth) lived next door. Marie described the burning of her grandmother's home: *I couldn't have been much more than six years old when my grandmother's house burned. It burned to the ground . . . right next door to us . . . I remember putting my nose up against the window pane and pulling away fast because it was so hot . . . There was no way of taking care of a fire, just the village men to come and help with a sort of bucket brigade. I know they lost everything, but nobody got hurt. We don't know how it got started. We never talked about it and they rebuilt it right away. Somehow the*

Inset: The Pooshee Plantation house as it appeared in 1939. Built in 1804, this home replaced the first one built in 1716. *Courtesy of the Library of Congress.*

Isabelle Deveaux Ravenel Macbeth built the Pooshee Plantation summer house in 1895. The first summer house, built by Dr. Rene Ravenel in 1856, burned in 1894.

lodge [where her grandmother taught village children] *behind the house didn't catch fire* (M. Ravenel 1991).

T. P. Ravenel's journal confirms the date of this fire, which burned his brother's house: *May 30, 1894, the dwelling house at Pinopolis of Mrs.* [Isabelle Ravenel] *Macbeth (Est. of Dr. Rene Ravenel) was burned.* Marie Ravenel's statement that the house was "rebuilt right away" indicates that the present Pooshee summer house was probably built in 1895.

The first chapter of Marie Ravenel's memoirs describes Pinopolis during her childhood. She spent most of her professional years as a missionary nurse in China and later retired to Charleston. Other families that have owned and have lived in the Pooshee summer house were the Binum M. Hudsons; Joe Heyward and his first wife, Julia Mitchell; and William and Alma Bright and their children, Billy and Midge. William E. Martin (1900–1987) and his wife, Bettina (1908–1997) modernized the house. Their children are William E. Martin Jr. (b. 1935) and Harriet (b. 1931).

Eugene and Jane Williams now own the Pooshee summer house.

Pooshee Plantation summer house living room in 2005.

Inset: Pooshee Plantation summer house dining room.

Rene Ravenel House — no longer existing

Morton Waring Lane (map: (T))

Walsh/Russell House

960 Morton Waring Lane (map: [28])

In 1879, Rene Ravenel married Maria Stevens Jervey. It is likely that they built this home around 1880. This was the childhood home of their daughter, Maria Ravenel, whose memoirs describe growing up in Pinopolis.

The Rene Ravenel lot is listed in Anne S. Ball's book, "Pinopolis: History of a Pineland Village," as the Walsh house. Ball did not associate this lot with Rene Ravenel and incorrectly said that the Rene Ravenel family lived in the Pooshee summer house. Maria Ravenel's description of the home where she grew up [Rene Ravenel house] is completely different from the floor plan of the Pooshee summer house: *The house in Pinopolis was built by my father* [Rene Jr.]. *It had a great big front porch. You entered the house on a hall with a room on each side . . . on either side of the dining room were large rooms, one for the boys and one for the girls. Upstairs there were two large rooms* (M. Ravenel 1991).

My father, Dr. J. N. Walsh, dismantled the Rene Ravenel house in 1941. My memory of the old house coincides with the description given by Maria Ravenel. In an interview with Elias Bull, Charles Cain confirmed that the Rene Ravenel house was located where the Walsh/Russell house is today: *Where the Walshes and Martins lived was built by the Pooshee father and son* [Dr. Rene Ravenel and his son, Rene] (Collected Papers of Elias Bull).

Rene Ravenel died in 1901. In his will, he conveyed all of his property to his wife, Maria Stevens Ravenel. On November 3, 1917, she conveyed her one-and-one-half-acre lot to Mary Causey, widow of former Berkeley County sheriff, Richard Greene Causey. Maria Stevens Ravenel then moved to Charleston with her children.

Mary Causey, her daughters, and one son moved to the former Rene Ravenel home. One of her daughters, also named Mary Causey, taught Henry R. Dwight Jr. (b. 1911) at the Pinopolis Academy. Dwight, who grew up in the Frederick A. Porcher house, remembers that Mary Causey had a brother, known as "Toots." Toots Causey drove Mr. Stroeker, a Moncks Corner pharmacist and Episcopal lay reader, to Trinity Church in Pinopolis every Sunday.

In 1939, Santee Cooper acquired this property in preparation for the creation of Lake Moultrie. Santee Cooper deeded the Causey property in 1941 to my grandfather, William Henry Sinkler (1872–1949) of Eutaw Plantation. My grandfather then deeded the property to my mother, Cleremonde Sinkler Walsh (1907–1958). My father, Dr. Joseph Norman Walsh (1901-1974), had the Rene Ravenel Jr. house, later known as the Causey house, dismantled. My father built the existing house on the old Ravenel/Causey site, which at this time consisted of thirteen acres.

The house that my father built has supporting beams that are about eighteen inches square, each made from a single longleaf pine tree. The beams, as well as the mantle, the heart pine flooring, the round porch columns, and the door hardware, were taken from my grandfather's home at Eutaw Plantation.

My father converted a large field on his property, running parallel to Lake Moultrie, into an airstrip. He subdivided the northern end of the property, including the airfield and pecan grove, in the 1960s. Today, the subdivision is entered from Pinopolis Road by Lucas Street, which leads to Walsh Drive.

The Percival Ravenel Porcher II house was built in 1901. This image was made in 1912. *Courtesy of Martha Porcher Ravenel.*

Dr. James Ravenel (1890–1984), son of Rene Ravenel who built the first home on this lot, told me that he (James) planted the three large white oak trees in the front yard of his family's home in 1900. In 1942, I planted the live oak tree that is near the southeastern corner of the present house.

At Dr. J. Norman Walsh's death in 1974, the property passed to his second wife, Sarah Edens Russell. She died in November 2002, and the property passed to her son, Otto Russell. He and his wife, Imogene, have restored the home, which is now their residence.

Percival Ravenel Porcher II House

901 Morton Waring Lane (map: 15)

Percival Ravenel Porcher II (1861–1931) owned Indian Field Plantation when he began construction of his Pinopolis summer house in April 1901. His uncle, John Henry Porcher, recorded that irons put in the chimney were 107 years old and came from the Santee Canal gates. He also recorded that the family moved into the summer house on June 29, 1901 (J. Porcher). Percival R. Porcher II and his wife, Martha "Mattie" Mitchell (1858–1932) of Charleston, had three children: Eudora Mitchell Porcher (1894–1903), Percival R. Porcher III (1897–1965), and Richard Dwight Porcher (1900–1973).

Ownership of the Percival R. Porcher house passed to Richard Dwight Porcher and his wife, Miriam Ellison (1906–1975), whose children are Richard Dwight Porcher Jr. (b. 1939), Martha Mitchell Porcher Ravenel (b. 1935), and Henry Ravenel Porcher (b. 1940). At the death of Dwight and Miriam Porcher in the 1970s, Robert LaFaye purchased the home and did extensive renovation, including enclosing the front porch. Elliott Williams Bishop now occupies the home.

The office of Dr. Morton Waring and Dr. Richard Y. Dwight is now on this property and is described with the Chelsea Plantation summer house.

The Percival Ravenel Porcher II house as it appeared in 2005.

Inset: Dining room of the Percival Ravenel Porcher house.

Ophir Plantation — no longer existing

Ophir Plantation, the ancestral home of the Porcher family, was six miles northwest of Pinopolis. Isaac Porcher, MD, the immigrant, arrived in Carolina in 1685. Isaac Porcher's birth date is unknown, but his family Bible is still in the Porcher family. Before his death in 1727, he either owned or had owned property at Jamestown Santee, Orange Quarter on the eastern branch of the Cooper River, the Goose Creek French settlement, and the upper waters of the Ashley River (Kirk).

Isaac Porcher's grandson, Peter Porcher (1726–1781) of Peru Plantation, bought Ophir before the American Revolution. Peter's son, Col. Thomas Porcher (1766–1836), inherited Ophir and built a home there about 1816. He was married first to Charlotte Mazyck and then to Elizabeth Sinkler Dubose (1788–1866). Between his two wives, Porcher had twenty-four children, fourteen of whom lived to maturity.

The Ophir Plantation house in 1901. *Courtesy of the South Carolina Historical Society.*

Ophir was the birthplace of the physician and botanist, Dr. Francis Peyre Porcher, grandson of Col. Thomas Porcher, and also great-grandson of the botanist Thomas Walter. Dr. Francis P. Porcher spent his boyhood and youth at Sarazins Plantation, the home of his parents, Dr. William Porcher (1800–1833) and Isabella Sarah Peyre (1803–1890).

In 1856, a cistern was built at Ophir that was about eight feet in diameter and situated about eight feet below ground level and six feet above. All of the bricks for the cistern were made on the plantation. A pump was used to deliver the collected rainwater into the house (Porcher and Haskell).

Ophir always remained in the Porcher family, and Henry F. Porcher (1891–1988) of Memphis owned it when flooded in 1941 (Kirk). This Henry F. Porcher was the son of John Couturier Porcher (1847–1911) and Eva Haskell. He was the grandson of the Henry F. Porcher mentioned next in regard to the Ophir pineland house.

Ophir Pineland House and Fishburne House

927 Fishburne Road (map: Ⓞ)

A 1901 photograph of Trinity Episcopal Church shows the Ophir summer house standing just northwest of the church. Henry Francis Porcher (1854–1903) and his wife, Margaret Deas Ravenel Porcher, along with his sister, Mary C. Porcher (1856–1926), were the last Porchers to occupy this home. After Henry's death from tuberculosis in 1903, his widow supported herself for a while by taking in boarders. She probably did this until about 1910 when she moved to the Woodboo summer house to live with her sister, Miss Bessie Ravenel, Pinopolis postmaster.

Ophir Plantation was the ancestral home of the Porcher family. Circa 1816, Col. Thomas Porcher built the home shown here. This image shows its appearance in 1939. *Courtesy of the South Caroliniana Library.*

One of Mrs. Porcher's boarders was Dr. William Kershaw Fishburne (1880–1968), who became a lifelong friend. Dr. Fishburne grew up in Walterboro, went to Porter Military Academy, and studied medicine at the Medical College of South Carolina. On the recommendation of Rev. William Wilson, MD, Dr. Fishburne moved to Pinopolis in May 1904 to begin his practice of medicine (J. Porcher).

After Dr. Fishburne and Anne Wickham Sinkler (1886–1981) were married in April 1910, they lived in the Ophir summer house. J. H. Porcher recorded: *May 8, 1910 . . . Dr. Kershaw Fishburne and his wife, Anne Sinkler, occupy house here owned by Miss Mary Porcher* [sister of Henry F. Porcher] *known as Ophir Pineland House.*

The Fishburnes later purchased the house and property and continued living in the summer house until their present home was completed in 1912. The

The Ophir pineland house as it appeared circa 1900. *Courtesy of the Cain files, South Carolina Historical Society.*

summer house was moved to the rear of the Fishburne house and used as part of a barn, which was taken down several years ago.

Anne Sinkler Fishburne grew up on Belvidere Plantation in upper St. John's. She cared passionately about the beauty of Pinopolis, and she was responsible for the planting of crape myrtle trees that were prominent along Pinopolis Road during my childhood, some of which remain to this day.

Dr. Fishburne began his practice using a horse and buggy for his rounds. He later used a motorcycle until the automobile became available. He is best known for his work with the Berkeley County Health Department. He became its director in 1930, during the worst of the Depression. He was well known for his efforts to improve the health of black residents of Berkeley County. His most famous associate was Maude Callen, a black nurse sponsored by the Episcopal Church to care for residents in the Pineville area. She practiced and taught midwifery. Because of her competence, compassion, and success, Callen was featured in *Life* magazine in 1951. This magazine article generated tremendous national interest, which brought about enough donations to create the Maude Callen Clinic in Pineville.

Dr. Fishburne, known as "Doc," was also instrumental in fundraising and organizing the construction of the Berkeley County Hospital in Moncks Corner. My father, Dr. J. Norman Walsh, Dr. Fishburne's nephew, became the hospital's first and only trained surgeon. The last administrator for Berkeley County Hospital was Moultrie Ball, Dr. Fishburne's son-in-law. In 1975, the Berkeley County Hospital was forced to close because it could not meet modern building standards for hospitals.

The Fishburnes built a clay tennis court, very close to the site of the original Ophir summer house, where they played and taught many village children the game. This tradition was still active in 1949 when Moultrie Ball taught me how to play on that same court. The court was abandoned when Moultrie Ball and his wife, Anne S. Ball, later built a court on Morton Waring Lane near their home.

When the present owners, Eugene and Virginia Koester, purchased the Fishburne house in 1983, the house needed extensive renovation. Gene Koester personally did most of the work. When I interviewed the Koesters in July 2005, Gene shared with me his experience of seeing a ghost in his new home. Gene and Virginia celebrated the completion of their house renovation in November 1985 and planned to move into it in December. They celebrated the occasion by having their family for Thanksgiving dinner. After dinner, the family was walking in the yard when they met Emily Fishburne Whaley, a daughter of Anne and Doc Fishburne. Emily was visiting Fannie Brawley, who lived next door. While in conversation in the Koester's yard, Emily Whaley casually remarked that her family home (the Fishburne house) had always been a "friendly" house. She then said, "If you meet a ghost, don't worry, it will be friendly."

Gene Koester thought nothing of the conversation until one night some time the following year. He was walking down the stairway when he looked down to the first floor and to the left, where he saw a lady sitting in a chair. She was sitting toward the front part of the chair with her back erect. Koester noted that she was dressed in a brown tweed jacket and that her hair was close to her head and wavy in the style of the 1930s. As soon as he took another step downward, she disappeared. He had neither met Anne Fishburne, who died in 1981, nor seen a photograph of her. When he described this experience to older

The William Kershaw Fishburne house replaced the Ophir pineland house in 1912.

people in Pinopolis, everyone confirmed that he had seen the ghost of Anne Fishburne.

Sarazins Plantation — no longer existing

See page 198 for a recently discovered sketch of the Sarazins plantation house.

F. M. Kirk's unpublished notes about Sarazins Plantation were provided by his daughter, Videau Kirk Simons. Kirk wrote: *The plantation west of Ophir derives its name from the first owner of whom there is a definite record, Jonathan Sarazin, who appears to have been a grandson of the emigrant Jean Sarazin (Sarazen, Sarrazin). The emigrant was a deacon in the Rev. Elias Prioleau's church at Pons, France, and followed his pastor to Carolina in 1687 . . . Jonathan and his brother, Moreau, were silversmiths, goldsmiths and engravers, operating their shop at the corner of Broad and Church streets in Charleston . . . under the sign, "The Tea Kettle and Lamp" . . . Early in 1800, Jonathan sold his St. John's holdings to Colonel Thomas Porcher of Ophir* (Kirk).

Thomas Porcher of Ophir gave his son, Dr. William Porcher (1800–1833), one thousand acres of land adjoining Ophir as a wedding gift when he married Isabella Peyre. Dr. William Porcher called it Sarazins and built a house there in 1825. The house was described in a letter written by Dr. Walter Peyre Porcher (1858–1919), grandson of Dr. William Porcher, to his sister, Virginia Leigh Porcher (1866–1940), as having a wide porch fifty feet long. There was a mahogany staircase with every fifth banister made of iron. The downstairs hallway went all the way through to the back porch where there was a garden with a view of the swamp. Dr. Walter Porcher noted that "Grandpa" William worked hard as a physician and also planted on a large scale. Dr. William Porcher's health was never strong. He died in Charleston in 1833.

"Grandma," Isabella Sarah Peyre Porcher (1803–1890), was left with six children under age ten. An unusual woman, Isabella chose to take over management of Sarazins Plantation and was later recognized as having managed better than most of the male planters of St. John's.

Unusual also was her determination to educate her children. When her children had outgrown home schooling by a governess at Sarazins, Isabella made plans to rent a house and move the entire family, nieces and nephews included, to Winnsboro. There, the children attended Mt. Zion Academy, run by a scholar known as "Old Hutson."

They carried all of their provisions for their new home in a caravan of wagons on a 130-mile trip to Winnsboro, which took several days. Walter P. Porcher continues the story: *I remember Grandmama telling me that one evening they lost their way and failed to get to the house they expected to spend the night in. It was raining hard and when quite dark they came to a small house, one of the party went in and asked if they could stop there—a poor tired anxious young woman declared that she had no room—that her baby was desperately ill and she had no time. Grandmama had not been married to a Doctor for ten years for nothing—Taking her little medicine chest with her, she went into the house where she found the baby in convulsions. In a short time—due to her capable directions—the child was quiet and sleeping. The mother declared that the Lord in heaven had sent her and of course could not do too much to make them all comfortable.*

Isabella's children became proficient in Greek and Latin. After the children's education was completed at Mt. Zion, Isabella returned to Sarazins. Having never lost her love of learning, Isabella started a circulating library for family and friends (W. Porcher).

F. M. Kirk also stated in his unpublished document about Sarazins: *When Chicora, or "New England City," boomed in the 1890s, Sarazins was sold by the heirs of Dr. William Porcher to the Emerson interests. When the tragic failure of that enterprise was realized, the plantation passed successively into the hands of various timber interests. It is now* [ca. 1939] *owned by West Virginia Pulp and Paper Company. Although the remains of the foundations of an older house are said to exist, it is doubtful that Jonathan Sarazin ever lived on the plantation named for him.*

Deserted of its inhabitants, the house [Sarazins] *gradually fell into decay. Finally the crumbling remains burned in 1924. Up until a few years ago, one could see the tall chimneys towering above the surrounding trees as one passed along the Congaree Road. They have since been torn down for the brick they contained.*

Some of the mantels of the Sarazins house are now in the present house at Bossis plantation on the Cooper River, and the stairway is in the home of Mrs. C. M. Wiggins in Moncks Corner.

In 1933, the WPA (Works Progress Administration) Camp was constructed on Sarazins plantation property and named Camp Francis Porcher in honor of Dr. Frances Peyre Porcher.

On the 1928 J. P. Gaillard plat, the J. F. Prettyman Lumber Co. is listed as owner of the Sarazins property.

Sarazins Plantation Summer House — no longer existing
(map:Ⓨ)

The Sarazins summer house may have been moved from Jack's Hole near Pooshee to Pinopolis.

Dr. Francis Peyre Porcher (1824–1895), one of two sons of Dr. William Porcher and Isabella Peyre Porcher, was born at Ophir but spent his childhood at Sarazins. He was first honor graduate of the 1847 class at the South Carolina Medical College. He studied medicine and literature for almost two years in Paris, France, and Florence, Italy. He had a successful private practice of medicine in Charleston and was a professor at the Medical College of South Carolina.

When the Civil War began, he enlisted as a surgeon in Holcombe's Legion. The southern army was dangerously short of medicines, and the Surgeon General of the Confederate States commissioned Porcher to write a medical botany book. In 1863, he published *Resources of the Southern Fields and Forests.* In this 610-page book, Porcher listed plants that had medicinal or economic value. It was widely used as a practical book for Confederate soldiers and civilians.

The Sarazins Plantation summer house was probably moved to Pinopolis from Jack's Hole circa 1840. *Courtesy of the Elias Bull Collection, Charleston County Library.*

A contemporary Cornell University botanist called it *the finest piece of research work of its kind in the literature of America* (Dwight).

After returning to his medical practice and teaching career in Charleston, Dr. Porcher was active in many local, national, and international medical organizations.

Julian Henry Porcher (1828–1908), the brother of Dr. Francis P. Porcher, also spent time at the Sarazins summer house. The last Porcher to own the home was Miss Anne Allston Porcher (1882–1970), daughter of Dr. Francis P. Porcher and his second wife, Margaret Bently Ward (1849–1895). Anne lived in Charleston and for many years was keeper of the Porcher family records.

Percival R. Porcher IV relates that he was born in the Sarazins summer house on July 5, 1931, while his mother, Marguerite Ravenel Porcher, was spending the summer in Pinopolis. The story was told to him that Dr. W. K. Fishburne and Dr. J. N. Walsh were at the Sarazins house for a July Fourth party when his mother went into labor. The two doctors remained to deliver him early the next morning.

In the 1940s, Mrs. Sarah Schuman, Miss Margaret Taylor, and Mrs. Ficken all shared the house identified as the Sarazins summer house while working for Santee Cooper. Margaret Taylor moved out after marrying John Tiencken, and Mrs. Ficken returned to Walterboro to live. Johnny Waddill remembers that Mrs. Schuman, his mother-in-law, was living at Sarazins in 1953 when he and his wife, Mary Ellen, moved to Pinopolis. When Mrs. Schuman moved into the Somerset summer house lodge in 1955, Johnny and Mary Ellen Waddill then moved into the Sarazins house. Every month, the Waddills mailed a check for $25 to the Sarazins summer house owner, Anne Porcher, who lived in Charleston.

Having moved from New York, the Waddills found living in Pinopolis to be an adventure: *In the winter in that old house I split lighter wood at dawn, built two fires in the double fireplace, lit the oil space heater, a kerosene heater, a gas heater, and turned on the electric stove oven and opened the door. Then I fixed breakfast while Mary Ellen tended to dressing Sarah and then went to work. It was wonderful.* The Waddills stayed in this home until about 1964 (interview with author, November 2005).

Later, Suzanne Parker and Jimmy Hill bought the house. They sold the property to N. D. Ford Jr., who razed the Sarazins summer house and built a new house on the lot.

Woodboo Plantation — no longer existing

Stephen Mazyck (1718–1770), twelfth son of Isaac the immigrant, built a home at Woodboo about 1740. F. M. Kirk records that there were so many Stephen Mazycks that they were given special names. At Woodboo was "Woodboo Stephen." His nephew, son of Benjamin Mazyck of Goose Creek, was called "Hickory-faced Stephen" Mazyck because of his rough features. "Upstairs Stephen" Mazyck of Charleston had such facial disfigurement from smallpox that he remained in his upstairs room for seven years. He accepted an invitation to visit Woodboo and, after that, would appear in public in Charleston (Kirk).

Woodboo passed to Stephen (b. 1749) and then to John W. Mazyck (b. 1754). John Mazyck sold it to Dr. Henry Ravenel (1790-1867) of Pooshee (Kirk "Plantations"). Thomas Porcher Ravenel, son of Dr. Henry Ravenel, recorded in his journal on Dec. 31, 1857: *During the past summer, the purchase and sale of Woodboo was agreed upon by father and John W. Mazyck as follows, Woodboo and Fairspring Plantations and the Pinopolis house and lot with the stock of cattle sheep and hogs which amounted to about 12 head of cattle, 18 head sheep, 10 grown hogs and few shoat to be delivered in Jan '58 for $12,000.* Thomas P. Ravenel also records that his father and the "girls" moved to the Pinopolis house on May 27, 1858.

At the death of Dr. Henry Ravenel, the Woodboo Plantation and summer house passed to his son, Thomas Porcher Ravenel. Thomas recorded on Nov. 12, 1867: *My house at Woodboo was burnt. Hot fire on the roof while we were at breakfast & before we perceived it, the garret was in a flame. We had no way of getting up to the fire & had to move out while we could. The only two men, except my two sons & myself, were out of hearing & we had no assistance but the alarm seemed to give all extraordinary strength & we saved more than we expected. Everything upstairs was lost except some bedding, which I drew through the fire that was then curling in flames through the garret door. Wind was very high & being a very old house, supposed to be about 170 years old, built of cypress, the house was a pile of ashes in 1 ½ hr. Before the house was entirely burnt, all of my former slaves from Fair-Spring & Wantoot were there rendering all the assistance they could. Moved my family over to Pooshee where Richard Dwight* [and] *my sisters were.* Thomas P. Ravenel remodeled the overseer's house at Woodboo and moved his family there in December 1867 (T. Ravenel "Journal").

Woodboo Plantation's first owner, Isaac Mazyck, purchased a sundial in England for his plantation home. His son, Stephen, brought the sundial to Woodboo. In 1939 when the plantation was to be flooded, this same sundial was moved to the front yard of the Woodboo summer house in Pinopolis (Kirk "Plantations"). Thomas Heyward Gignilliat, great-grandson of Thomas P. Ravenel, says the sundial is now at Cypress Creek Plantation in Jasper County near Tillman, South Carolina (interview with the author, October 2005).

When Santee Cooper flooded Middle St. John's, Woodboo was still in the Ravenel family.

Woodboo Plantation Summer House

no longer existing

Dooley Lane (map: S)

When Dr. Henry Ravenel purchased Woodboo Plantation from John Mazyck in 1857, the summer house must have been part of the bargain since his son, Thomas P. Ravenel, recorded in his diary: *May 27, 1858, Father and the girls move to Pinopolis into the house he bought from J. W. Mazyck.* John W. Mazyck had a racetrack on this lot in 1853, which was described by Thomas in a letter to his brother, Rene, in Paris: *John Mazyck is as usual. He is racehorse mad. He has a track around his lot & is to have three horses in training this fall . . . Col. Mazyck will be an impatient man come the next races in Pineville, as there will not be one of Mr. Sinkler's* [W. H. Sinkler was the author's great-great-grandfather] *stable in training . . . already it is amusing* (T. Ravenel "Journal").

The Woodboo summer house was probably built by "Woodboo Stephen" Mazyck and moved from Jack's Hole (near Pineville) to the Woodboo Pinopolis lot in 1846. John Henry Porcher noted that Uncle Samuel Porcher (1768–1851), who owned Mexico Plantation, gave the house to John Henry Porcher's half-brother, Isaac Porcher V (1812–1849) (J. Porcher).

This image shows the Woodboo Plantation summer house as it appeared before 1908. It was probably moved to Pinopolis from Jack's Hole in 1846. Vandals burned it in 1981. *Courtesy of Martha Porcher Ravenel.*

The house passed to Dr. Henry Ravenel's son and then to his grandson, Thomas P. Ravenel Jr. (1850–1922), who lived in Savannah. Thomas P. Ravenel Jr. left the summer house to his two sisters, Elizabeth Ravenel (1852–1935) and Margaret Deas Ravenel Porcher (1857–1944), who married Henry Francis Porcher of Ophir. Deas Ravenel Porcher lived at the Ophir summer house until her husband died; then, she lived with her sister Elizabeth Ravenel, known as Bessie, at the Woodboo summer house. J. H. Porcher recorded in 1908 that *Bessie commenced today to put upper story to her house.* Dr. James Ravenel further confirmed that between 1908 and 1910 a second story was added to the Woodboo summer house. This is documented by his note on a photograph in the possession of Martha Porcher Ravenel.

Bessie and Deas' brother Heyward Ravenel (1868–1903) married Marguerite Hamilton Seabrook, who died a few years after the birth of their daughter Marguerite. Bessie and Deas Ravenel raised Marguerite at the Woodboo summer house in Pinopolis. Marguerite married Percival Ravenel Porcher III and later lived in the Chapel Hill summer house.

The Woodboo house was sold to Lewis and Rebecca Nexsen, but vandals burned it in 1981 while the Nexsens were restoring it. They have built another house on the same location.

This image of the Woodboo Plantation summer house, made in 1970, shows the second story, which Elizabeth Ravenel, the Pinopolis postmaster, added in 1908. *Courtesy of the Elias Bull Collection, Charleston County Library.*

Inset: This desk sat on the front porch of the Woodboo summer house and was used for mail storage before the Post Office building was completed. It is now in the home of Bertha Cain in Mount Pleasant.

Old Post Office

1940 Pinopolis Road (map: 26)

A description of the early postal service in Pinopolis comes from Emily Guerard Ravenel of Savannah (b. 1890) in a letter she sent to Florence Lucas, who was living in Pinopolis. Emily was the granddaughter of Thomas Porcher Ravenel and the niece of Elizabeth Catherine Porcher Ravenel. She wrote, *My first recollection of the Post Office was before . . . 1899. Daddy Isaac* [Hector] *Myers . . . used to walk the mail from Moncks Corner with a sack on his back and sit on the steps while Grandpa . . . sorted the mail . . . The post office was a desk on the piazza between the windows of the living room* [of the Woodboo summer house]. *It was a drop front desk* (E. Ravenel).

The old post office, which stands a few feet east of the present post office, opened for the first time on May 16, 1894 (T. Ravenel "Journal"). The first postmaster was Miss Elizabeth (Bessie or Aunt Bet) Catherine Porcher Ravenel (1852–1935). She lived in the Woodboo summer house, just east of the old post office. She was the daughter of Thomas Porcher Ravenel (1824–1898) and Elizabeth Margaret Wilson (b. 1827).

Emily's letter continues: *Between 1910 and 1916, Aunt Bet had to stand a Civil Service Examination to hold her position as Post Mistress, and she was sent information as to how to prepare for this exam . . . It was amazing how well prepared she was when you think she left school at fourteen . . . Bet went to Washington, D.C., as the oldest post master in the U.S. in length of service* [1894–1935] *. . . I don't know when the Lodge* [the old post office] *was built, but I think Mr. De Veaux* [Stephen Deveaux] *must have come to live in it . . . and help in the Post Office . . The Pinopolis mail was then sorted into approximately twenty-five stacks to be picked up by the families . . . Only the front room was used as the Post Office. The rear room served as a schoolroom and residence for Mr. De Veaux who taught many of the youngsters of the village.* Emily Ravenel's last statement confirms that at one time the post office was used as a school. According to Emma Jenkins Gribbin, Stephen Deveaux lived in the house next door to the post office with Bessie Ravenel (Gribbin 1968).

In the early years of the post office's operation, office hours were 9:00 a.m. to 1:00 p.m. It was possible to get mail and stamps at other times because Bessie Ravenel had a bell placed between the post office and her home so that customers could ring for service. The original bell is still in place.

Marie Ravenel (1892–1991) described how important the post office was in the social life of the villagers: *When summertime came, we'd gather at the post office in the morning . . . the post office was in Aunt Bet's yard . . . it was a small building to the side of their home in the center of the village. We'd gather there on the outside and make the plans for the day. The boys were looking for a swimming stream . . . the girls would plan on playing cards at one house all morning. Lizzie Lucas had a nice little reading club for the girls once a week and we'd read different books . . . she'd read aloud. The girls never went swimming because there were no swimming suits and the boys all went in the raw. We girls were absolutely forbidden to visit the stream* (M. Ravenel 1991, 18).

The adults also used the post office as a gathering place, a tradition still active when Suzanne Hill was postmaster over one hundred years later. She served from 1972 until 2001. Suzanne especially remembers the visits of Anne Fishburne, Sarah Schuman, Kinloch Bull, Julia Kirk, and Margaret Stevens, who were the oldest citizens of Pinopolis and the last of the generation who were part of the "old era." They met almost daily at the post office. Calling them "wonderful spirits," Suzanne describes them as the kindest and gentlest people she has ever known.

The old Pinopolis Post Office began service in 1894.

Suzanne Hill remembers that when she became acting postmaster in 1972, she would arrive at the old building at 6:00 a.m., turn on the single light bulb that hung from the ceiling by its cord, and then raise the American flag. In winter, Henry Lucas would arrive at 6:15 a.m., bringing wood, and start a fire in the old stove. He would come back about 2:00 p.m. to check on it. Charles Cain brought her coffee every morning.

There were no mail boxes, just a shelf where Suzanne put customers' names on with tape. As residents came in, Suzanne would personally hand them their mail. When customers registered to have a place on the shelf, she would ask for their dog's name and put it on the registration card. It wasn't unusual for a customer to call Suzanne at the post office in search of his dog.

When Suzanne Hill lived near the Pinopolis Methodist Church, neighborhood dogs followed her car every day on her way to work at the post office. Other dogs joined in along the way. By the time she got to the post office, there were as many as four dogs ready to come in as soon as she opened the door. After turning on the light and raising the flag, she gave each dog a biscuit from her pocket. Then all the dogs would return to their homes for the day. The ritual began again the next morning.

Suzanne recalls that Wayne Creech had a beagle named Toby who would come and sit with her on the days he could escape from his electric fence. This only happened when the battery on his electric collar was discharged. Andrew Dennis, son of Markley and Jan Dennis, had a chocolate Labrador retriever named Hank, who would follow Andrew from his home (the Somerset summer house) to the school bus stop. After accompanying Andrew to the bus stop, Hank would then come in and sit with Suzanne until the bus came back at 2:30 p.m. Hank knew the sound of the bus, and as soon as it arrived, he would leave the post office to meet Andrew. Suzanne remembers the sad day when someone poisoned Hank with antifreeze, and he came to her for help, but she could not save him. Suzanne misses Hank's company and recalls the sadness and loneliness without him (interview by author, July 2005).

The Charleston *Post & Courier* published a story about Suzanne and the post office on January 7, 1993, describing how Suzanne knew all her customers by their first names. Suzanne said she felt that every customer was just a member of her family. Some residents referred to her as "the local newspaper." Suzanne was in charge when the old post office building was put out of service in 1975 and the new post office built.

The following table comes from the U.S. Postal Service's records of Pinopolis.

Name	*Title*	*Date Appointed*
Elizabeth P. Ravenel	Postmaster	02-10-1894
Mrs. Elizabeth R. Cain	Acting Postmaster	11-07-1935
Miss Katherine M. Cain	Postmaster	06-03-1936
Mrs. Vera Williams	Acting Postmaster	11-17-1941
Mrs. Vera Williams	Postmaster	04-28-1942
Mrs. Lillian W. Skinner Pace	Acting Postmaster	08-09-1950
Mrs. Lillian W. Skinner Pace	Postmaster	02-29-1951
Mrs. Suzanne Parker Hill	Officer-In-Charge	06-30-1972
Suzanne P. Hill	Postmaster	04-26-1975
Nancy E. Knight	Officer-In-Charge	07-02-2001
Nancy E. Knight	Postmaster	06-28-2003

On December 19, 1990, the Berkeley County Historical Society received title to the Old Post Office from the Cain family. The old post office building has significant problems with deterioration, and a contractor estimates that restoration will require $30,000. Proceeds from the sale of this book have been

This image of the old Pinopolis Post Office shows that windows were added after 1924.

Inset: Old Post Office in 1924, left to right: Bertha, Maude, and Mary Dwight (Davis). Note the absence of windows. *Courtesy of Katherine Davis de Loach.*

The old Pinopolis Post Office, ca. 1901, with Stephen Deveaux, who taught students in the room to the rear of the building. *Courtesy of the Cain files, South Carolina Historical Society.*

This letter to Mrs. M. L. Macbeth was postmarked first in Charleston and again when it arrived in Pinopolis the following day, April 14, 1905. *Courtesy of the Berkeley Museum.*

assigned to the Berkeley County Historical Society. A portion of those funds will be used to restore the old Pinopolis Post Office. Tax-deductible contributions may also be made to the Berkeley County Historical Society for the Pinopolis Post Office Restoration and Maintenance Fund, BCHS, P.O. Box 504, Moncks Corner, SC 29461. Since the first printing, it has been discovered that the Old Post Office was built in the 1850s as the Woodboo summer house lodge. It is likely that Steven Deveaux lived there and used both rooms before it became the post office.

Wampee Plantation
(map: 38)

The Wampee Plantation house is the only Middle St. John's plantation house still standing on its original site. Two other plantation houses in Middle St. John's were moved to other locations at the time of the Santee Cooper project: the Hanover Plantation house to Clemson, South Carolina; and the Woodlawn Plantation house to Georgetown, South Carolina. Today, the Wampee Plantation house is considered part of Pinopolis, but no one thought of Wampee as being part of Pinopolis Village before Santee Cooper flooded the area.

Wampee Plantation has been the subject of a great deal of research over the years, but inconsistencies have prevented giving a totally accurate picture of the plantation's history. Names of early owners, as well as dates of ownership, vary from resource to resource. The timeline presented below is a compilation of materials collected from Berkeley County historians M. C. Orvin, F. M. Kirk, Russell Cross, Elias Bull, and Anne S. Ball.

1696 John Stewart received a grant from the Lords Proprietors for 804 acres, some of which became a part of Wampee.

1698 Rev. William Screven, the first Baptist minister to come to Carolina, bought 804 acres in the Wampee area from Stewart on January 23, 1698.

1700 Screven purchased 300 more adjacent acres, built a dwelling house, and named it Somerton for his birthplace in Somersetshire, England. Thirty Baptist families lived in the area, but there is no known record of where they lived or where they might have worshipped.

1701 Mrs. Elizabeth Wetherick, sister to Mrs. William Screven, bought 325 acres of her brother-in-law's property south of Somerton and named her plantation Wampee. This transaction is said to be the first naming of what is now Wampee Plantation (E. Bull 1962).

1704 Screven's home at Somerton burned. He moved to Charleston.

1708 Screven sold his original 804 acres in the Wampee area, plus the Stewart tract of 300 acres, to Rene Ravenel, a devout French Huguenot.

1728 Wampee was sold to Paul Ravenel, who built a home on the site of the present-day Wampee Plantation house (E. Bull 1962).

1749 Gabriel Guignard owned Wampee and lived there, either building a home or using the Ravenel home.

1768 Elizabeth Clevland bought Wampee from James and Ester Keith and made Wampee her home.

1771 In April, Elizabeth Clevland married Dr. William Hardcastle, who joined her at Wampee. F. M. Kirk ("Plantations") mentions that Dr. Hardcastle lived at Wampee and that at his death, his widow moved to St. Stephen. Kirk does not mention, however, that Elizabeth Hardcastle was Wampee's owner or that she was "a woman of color." The remarkable story of Elizabeth Clevland Hardcastle follows at the end of the Wampee story.

1771 In June, Elizabeth Clevland Hardcastle sold Wampee to Gilbert Ross and James Mills, London merchants.

1772 Ross and Mills sold Wampee to Thomas Sabb.

1790 Charles Johnston purchased Wampee. Charles's daughter Catherine inherited it.

1798 Catherine Johnston married James Macbeth. They had four children: Mary Macbeth (1800–1865), who

Constructed between 1822 and 1844 by Charles Macbeth, this is the third or fourth Wampee Plantation house. This image was made ca. 1939. *Courtesy of the South Caroliniana Library.*

married William Robertson; Robert Macbeth (1802–1867); Charles Macbeth (1805–1881), who was mayor of Charleston from 1857 to 1865; and James Macbeth (1809–1821). Charles Cain, in an interview with Elias Bull, said that Mary and William Robertson owned Wampee before it was passed to Mary's brother Charles Macbeth.

c. 1822–1844 Charles Macbeth built a home (the existing building) for his mother, Catherine Johnston Macbeth.

Charles Macbeth and his wife, Henrietta Gourdin Ravenel, left Wampee to their son, Richard Yeadon Macbeth (1842–1903). Richard Yeadon Macbeth's widow, Mary Louisa "Aunt Lou" Gaillard Macbeth (1851–1913), next owned Wampee, but having no heirs, she left Wampee to her husband's nephew, William Cain (1874–1926). William Cain was the son of Dr. Joseph Palmer Cain and Mary Catherine Macbeth, sister of Richard Yeadon Macbeth. William Cain already owned Somerton Plantation and Somerset Plantation. After inheriting Wampee Plantation from his Aunt Lou Macbeth, Cain's landholdings totaled about six thousand acres (W. Cain). William Cain's son, William Cain (1908–1993) of Darlington, noted that after the death of his great-aunt Mary Louisa Macbeth in 1913, the Wampee house was never used again (W. Cain).

1939 Santee Cooper acquired Wampee Plantation from the Cain family by condemnation.

1944 Santee Cooper renovated the Wampee Plantation house and began using it as a retreat for visiting state dignitaries and local officials.

1945 The Cain family sued Santee Cooper for the return of that part of Wampee that was not essential for the Lake Moultrie project. The South Carolina State Supreme Court returned 33.3 acres of Wampee Plantation land that had not been flooded, which included Wampee Plantation house, to the Cain family. The Cain family later sold the same land back to Santee Cooper.

1988 Santee Cooper made Wampee Plantation part of its conference center. It continues in this function in 2006.

Wampee, during pre–Revolutionary War times, referred to an area extending from Biggin Church to present-day Wampee Plantation house in Pinopolis. There are several competing ideas as to how Wampee got its name. A widely accepted explanation (W. Ball 1983) is that the area was named for the Indian chickweed "Wampee" that grew in abundance. Local botanist, Dr. Richard Porcher, finds no evidence to support this claim.

Another explanation is proposed by longtime Pinopolis resident, William "Bill" Wiley (1922–1996). Wiley notes that the Etiwan Indians first occupied St. John's Parish and that the Cooper River was initially called the "Etiwan River." The Etiwans had a trading post near the headwaters of the river called

"Wampee." The Etiwans did not accept the traders' European money. Instead, they accepted only shells and trinkets known as "wampum" (Wiley 1968).

The present-day Wampee Plantation house is at least the third structure built on that site and could even be the fourth. F. M. Kirk stated in 1939 that traces of two previous homes could still be found on the Wampee site. Kirk explained that during construction of the present Wampee home, a tornado broke the framing near the sills. The workers just cut off the broken parts, and the result was a house smaller than planned.

During my childhood and early teen years, the Wampee Plantation house was unoccupied, unlocked, and seemingly in ruins. It was literally a haunt for my friends and me, but it was a somewhat dangerous place to explore. We were always barefooted, and broken glass was all over the floors. The doors creaked when we moved them, and the light inside was dim.

I was not aware of any ghost stories about Wampee when I was growing up in Pinopolis. This changed, though, when I was about twelve years old. On one exploratory visit with my friends, I led the way onto the porch and opened the door. I was amazed to see a woman's figure on the stairway landing, between the first and second floors. The figure stood near a window on the stairway landing. She had a pale, indistinct face and was wearing a transparent, chiffon-

The Wampee Plantation house in 2005.

Inset: Arrowheads found at Wampee Plantation. *Courtesy of James B. Skinner.*

like, flowing garment. I don't remember being afraid; however, my friends and I left in a hurry, and I never went back. William Cain noted in his memoirs that Wampee had a *community-wide reputation for being haunted; and innumerable tales were told of queer apparitions, unexplained noises and actions, and the like, many of them experienced by different people at different times.*

Wampee Plantation's very early history is rich in Indian fact and legend. William "Bill" Wiley became a local expert on the Etiwan Indians living in the area. In 1939, Wiley, along with his high school chemistry class, took part in the excavation of Indian burial mounds, or tumuli, found on Wampee Plantation. Evidence found in the tumuli indicated that the Indians laid to rest there had probably been part of a battle between Etiwans and white settlers in the area. Positioning of the bones confirms that the Indians had suffered trauma. Both male and female bones were found.

In 1955, Wiley wrote about this excavation at Wampee in a paper called "Tumuli in the Pinopolis Reservoir." Wiley observed: *Within a few hundred yards and directly South of this Indian tumulus, across a very thick growth of trees and underbrush, is the old long-vacant Wampee Plantation dwelling which has the distinction of being listed among the haunted houses within the Pinopolis area. Shall we, like some of the inhabitants of the neighborhood, believe that the following ghost story is evidence of the reincarnation of one of the young squaws who followed her mate into battle and met the sad fate as described above?*

Wiley continued: *Legend tells us that on rainy nights just before low twelve, when the bats become vigilant in the attic, when the owls begin to hoot in the ancient oaks around and the snakes leave their bed of concealment to crawl about, then the ghost appears upon the front steps of the Wampee building, clad, not in the yellow skin of the fawn, but in a more ethereal garb, a flowing robe resembling blue silk chiffon with Cinderella-like slippers on her feet and a complexion as exquisite and delicate as a blush of early dawn upon the morning dew; it makes its usual visits to all the rooms of the house and to the surrounding premises; then the ghost vanishes as mysteriously as it appeared . . . This story is known and repeated by many of the older and most highly respected citizens in this vicinity, and there has never been a question as to its reality by anyone who claims to have seen the apparition* (Wiley 1961).

Wiley's paper on Indian tumuli allowed him to pay tribute to the earliest inhabitants of Wampee and to ponder their fates: *From Ophir Plantation, the Etiwans evidently continued their retreat westward and established a new frontier across the Wassamassaw, which goes beyond the confines of the Pinopolis reservoir* [Lake Moultrie]. *Eventually the Etiwans, as well as practically all other Indian tribes in the low country, were driven out or exterminated. We in America who now condemn aggressor nations across the sea should remember that our forebears also were aggressors when they ruthlessly drove out and destroyed this native people. What a pity there is no one left among the race of Hiawatha, and no hand to lay a wreath upon these graves. There is no indication that there will ever be erected a shaft perpetuating the memory of the Etiwans who fought and died to save their own homeland* (Wiley 1961).

Elizabeth Clevland Hardcastle (1741–1808)

Elizabeth Clevland Hardcastle was mentioned briefly in connection with the history of Wampee Plantation, as she was its owner in 1771. Her story is an unusual one because of her racial background. Her legal status and, just as importantly, her social status in the state of South Carolina were greatly affected by Hardcastle's racial identification as "a woman of color." Author and historian E. Louise of Columbia, South Carolina, spent some twenty years researching Elizabeth Hardcastle and wrote a book about her called *Elizabeth Clevland Hardcastle, 1741–1808: A Lady of Color in the South Carolina Low Country* (Louise 2001). Much of the information related in this book about Pinopolis and its plantations comes from Louise's book.

Born in the Banana Islands near Sierra Leone, West Africa, Elizabeth's English father was Capt. William Clevland (1720–1758) of the Royal Navy. Her grandfather (1665–1735), with the same name and rank, was a distinguished citizen of Scotland.

Elizabeth's father was shipwrecked near the Banana Islands. He was rescued but decided to remain and to make his home there. Clevland worked for an African king, Chief Skinner Corker, who had been educated in England and whose father was an Englishman. Captain Clevland married Chief Corker's daughter Kate and became a slave trader. Captain Clevland and Kate had two children, John (1740–1771) and Elizabeth (1741–1808). Both children were schooled in England.

Elizabeth was taken on a visit to Carolina at about age six by her father. The party visited Dr. William Keith of Keithfield Plantation near Moncks Corner. Frederick A. Porcher wrote in his memoirs that his informants reported that Elizabeth wore the finest dresses and accessories and was envied by all the little white girls from the nearby plantations.

Elizabeth returned to Carolina again in 1764 at age twenty-three, accompanied by Catherine Clevland, the daughter of her brother John. Elizabeth came with the intention of buying property, becoming a citizen, and raising her niece. They arrived in the colony so early in the history of South Carolina that there were no laws yet governing the legal status of persons of mixed ancestry or mixed race. Free persons of mixed color or race abided by the same general rules as those of the predominant racial and social class. South Carolina's views on racial standing changed significantly after the Revolutionary War, though, and by 1803, free people of color were denied rights such as holding religious services or any meeting between sunset and sunrise. Elizabeth's opportunities and freedoms were called into question because of her racial heritage even though she was free and not a slave.

The French Huguenots, who themselves had come to Carolina to escape religious persecution, were mostly welcoming, friendly, and helpful to Elizabeth and Catherine as the women attempted to make a place for themselves in the Low Country. Elizabeth was classified as "white" in the U.S. Federal Census of 1790 and 1800. Even so, because of her English father and African mother, Elizabeth had to carry legal papers stating that she was "free." Elizabeth lived among the planters of St. John's and was accepted as an equal by most. In his memoirs, Frederick A. Porcher calls Elizabeth a mulatto and expresses his criticism of those families who chose to receive Elizabeth into their homes: *Her father* [Captain William Clevland] *had connections in the colony, and must have been a man of great influence here . . .* [Her father] *succeeded in getting her under the protection of some of the principal families of the colony . . . it is notorious that she associated on terms of social equality with her neighbors . . . and gave dinner parties, and was invited to dinner parties. My informant* [Mr. Dubose] *had more than once been one of her guests* (F. Porcher 1943).

In 1768, Elizabeth Clevland purchased Wampee and Brick House plantations and in 1771 married Dr. William Hardcastle, an English physician. In the same year of her marriage, she sold Wampee Plantation. Dr. Hardcastle died in 1777, possibly while fighting as a patriot during the Revolutionary War. Before Dr. Hardcastle's death, he had practiced medicine and had business interests in St. John's and St. Stephen.

Before Dr. Hardcastle died in 1777, he had trained his wife to handle simple medical problems and some surgical procedures. Elizabeth had become quite proficient. F. A. Porcher related personal information given to him by Samuel Dubose: [My] *grandfather died there in her house, having gone there for the benefit of her surgical skill; for she had the reputation of having learned a great many of the secrets of surgery from her husband and had a reputation for skill which gave her a large practice.*

In 1780, Elizabeth Hardcastle purchased Raccoon Hill Plantation from Joseph and Catherine Palmer. She lived there and capably managed all aspects of plantation work and life. She owned cattle and cultivated cotton and rice. She also owned slaves, who were vital in providing the labor force for cotton and rice production. Elizabeth was a statistical oddity for her time. In South Carolina in 1790, there were only 125 free Negroes. Seven of those free Negroes owned a total of twenty slaves. In the United States shortly before 1865, free Negroes owned 18,000 slaves.

In 1797, Elizabeth Hardcastle purchased Pierce's Old Field Plantation near Pineville. Pineville was settled in 1794, but there was no evidence that Elizabeth had a summer home there. In 1804, she purchased Tucker's Plantation from Hugh Grant, who had purchased it from the estate of Peter Sinkler. This property was just south of the St. Stephen Episcopal Church. Elizabeth moved to Tucker's Plantation with her niece Catherine, who by this time had a family of her own.

Although Elizabeth Hardcastle had all the outward trappings of the planter class to which she aspired, her social position was a precarious one. As F. A. Porcher relates in his memoirs, *Her position in the parish was a curious and anomalous one. Many of the people asserted the prejudice of race, and would hold no intercourse with her; among these was my father's family; others on the other hand visited her and were visited by her; and among these were at least some of my mother's family. She lived in Fairforest Swamp near my granduncle Thomas Palmer, and Mr. Frank Peyre, and as she had a considerable estate and no children, these gentlemen hoped by their attention to be paid for their compliance in her will. I have heard Mr. Dubose say that he had eaten many a good dinner in her house, in company with the Peyre family, themselves conspicuous for their piety or their morals. The last of the regular rectors* [early 1800s] *of St. Stephen's church was an Irishman who lived notoriously as the paramour of Mrs. Hardcastle (F. Porcher 1943).*

After Hardcastle's death in 1808, there was pandemonium. She left a will, but it was burned by one thought to be her friend, whose real objective was to obtain her property. Elizabeth was deemed to have died intestate, and her considerable estate was attacked by the greedy from all sides. Her estate was in the courts from 1808 until 1825.

Even though Elizabeth's niece Catherine Clevland, who was also known as Free Kate, had been given a legal statement of ownership of Raccoon Hill Plantation at her aunt's death, the courts contrived to deprive Kate Clevland of this property. South Carolina law stated that free persons of color could not testify against a white person in court. Therefore, Catherine Clevland was powerless to give witness or share her personal knowledge about her aunt's will in a court of law. Eventually, Kate got Raccoon Hill; however, her other inheritances had been squandered. By 1822, free persons of color were required to have a white guardian and were required to register with the courts twice a year. A tax was imposed on all free persons of color. Any free person of color, who had left the state earlier, could not reenter.

Catherine Clevland, or Free Kate, lived until 1845, dying at age one hundred. Although she was baptized at Biggin Church in 1766 as Catherine Clevland, the South Carolina legal system would recognize her only as "Free Kate." E. Louise, author of the book about Elizabeth Clevland Hardcastle, is descended from Catherine Clevland's eldest daughter, Rebecca, who died in 1849. In 1941, before being evicted by Santee Cooper, some of Catherine Clevland's descendents still owned property in the Raccoon Hill vicinity (Louise 2001).

A 1928 map of Berkeley County by J. P. Gaillard (below) shows that Kate Clevland's property was still recognized. Kate's property is highlighted in the center of the map.

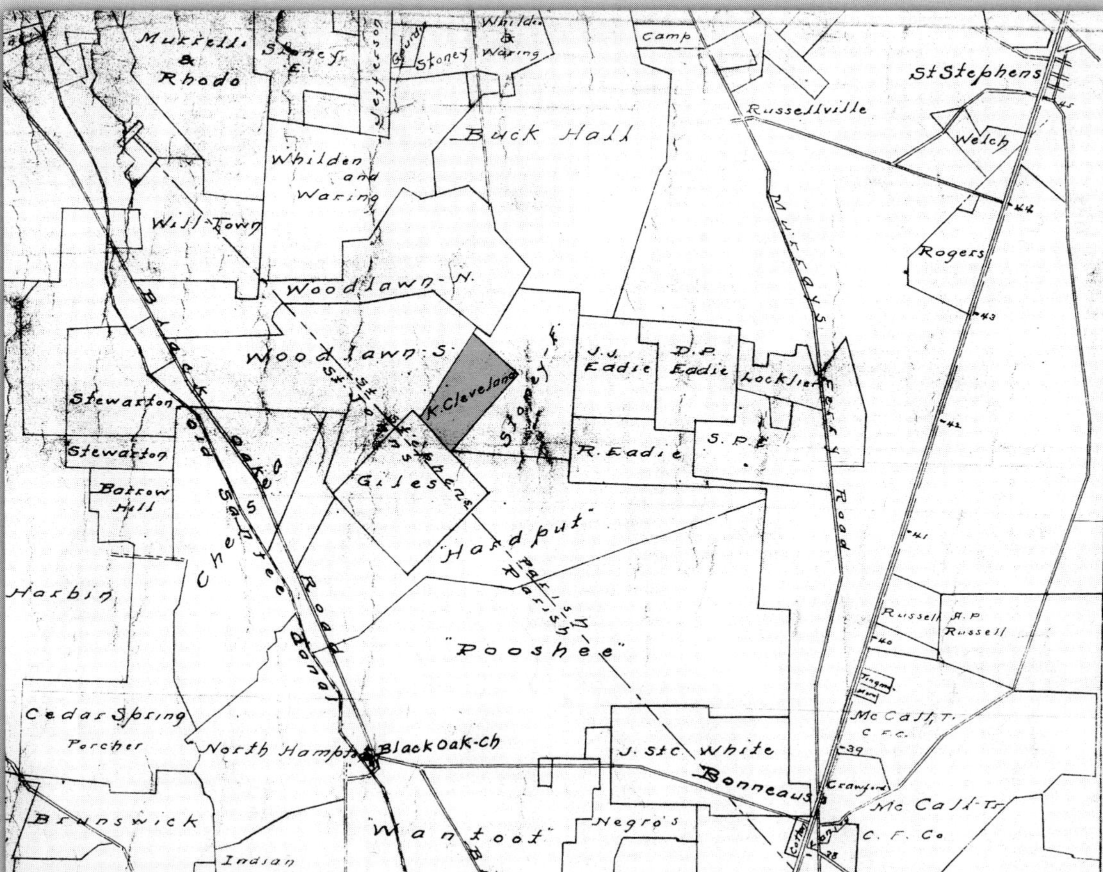

Wampee Plantation Summer House

2005 Pinopolis Road (map: 29)

Charles Macbeth built the Wampee Plantation summer house sometime between 1822 and 1844. Anne S. Ball writes that Macbeth built the Wampee Plantation summer house in 1836, but this is not documented. In Elias Bull's interview with Charles Cain about the Wampee summer house, Cain stated that the west wing of the summer house was first part of a Sullivan's Island home that Charles Macbeth owned. Macbeth served as mayor of Charleston during the Civil War. When it became unsafe to remain on Sullivan's Island during the Civil War, the Honorable Charles Macbeth had a portion of the island house shipped by boat up the Cooper River and then transported over land to Pinopolis. It is possible, also, that the Wampee summer house was first built elsewhere and then moved to its Pinopolis location. Elias Bull reported that before 1834 some homes from Raccoon Hill near Pineville were dismantled and moved to Pinopolis (E. Bull 1965).

The Wampee Plantation summer house was built between 1836 and 1844. It may have been moved here from Raccoon Hill near Pineville. This image was made in 1970. *Courtesy of the Elias Bull Collection, Charleston County Library.*

I examined the floorboards in the main living room of the Wampee summer house in 2005 and noted that they were numbered. Numbering boards was a common practice used to ensure correct ordering for dismantling lumber and rebuilding on a second site. It is possible that the main part of the Wampee summer house was moved from Raccoon Hill and reassembled at its present-day location.

Charles Cain said in his interview with Elias Bull that Lou Macbeth, daughter-in-law of the original builder Charles Macbeth, added a second story to the summer house. John Henry Porcher confirmed in his diary that on January 15, 1907, Lou Macbeth was adding a second story (J. Porcher).

In the 1950s, Elizabeth "Lill" Ravenel Cain and her husband, L. Whitaker "Whit" Boykin, occupied the Wampee summer house along with Lill's brother, Charles Cain. When the Boykins moved from Pinopolis, Charles Cain (1910–1981) continued to live in the Wampee summer house until he died. After Charles's death, Whit and Lill Boykin's daughter, Marcia Boykin Spainhour, and her husband, Robert, bought the house from the Cain heirs. The house was still primitive at that time; however, the Spainhours succeeded in modernizing the interior without destroying the house's charm or significantly altering its appearance.

The present owners of the Wampee Plantation summer house are Jane and Verne Pulling.

This image of the Wampee summer house in 2005 shows the west wing, which was moved by barge in 1860 from Charles Macbeth's Sullivan's Island beach home.

Somerset Plantation — no longer existing

The first reference to Somerset Plantation is found in a description of a land grant for 804 acres to John Stewart in 1696. Two years later, Stewart sold the acreage to Rev. William Screven, the first Baptist minister to South Carolina. In 1700, Screven added 300 acres to the original tract and built his home there, hoping, perhaps, to establish a Baptist community. Screven named his home Somerton. Wampee, Somerton, and Somerset plantations all stem from this original purchase by Screven.

Rev. Screven's property increased as acres were added, but in 1704, Screven sold his St. John's properties to Rene Ravenel. Somerset is referred to by name for the first time in 1736 when Paul Ravenel, son of Rene Ravenel, conveyed 725 acres of land called Somerset to his brother, Daniel Ravenel (Kirk, "Plantations"). Somerset Plantation then passed through Ravenel and Mazyck families until Nathaniel McCants purchased it in 1811 (Caroline Cain 1958).

At this time, there were nine tracts of land associated with Somerset Plantation, consisting of 1,463 acres. William Cain (1792–1878) and Isaac Dwight purchased the property in 1827. In 1830, Dwight sold his portion to Cain. By 1840, William Cain had increased the property to 2,175 acres and had remodeled a house, built earlier by Benjamin Mazyck, to function as the Somerset Plantation dwelling house. Somerset Plantation house became the seat of the Cain family until it was demolished by Santee Cooper to make way for Lake Moultrie in 1940 (Caroline Cain 1958).

In 1939, F. M. Kirk wrote that Somerset was one of the few plantation homes in Middle St. John's Berkeley still inhabited by the family of its antebellum owner. Kirk described Somerset Plantation house as still being in good repair and still being the seat of that hospitality which made the South famous in a happier day.

In an address to the St. John's Hunting Club in 1907, Rev. Robert Wilson, MD, featured Somerset's owner William Cain: *The last of these commanding figures*

The Somerset summer house described on page 91 is the only remaining historic home in Pinopolis that fits the typical architectural pattern of its time. Margaret Stevens in "Pinopolis Village," *Names in South Carolina*, Winter 1982, described this pattern as follows:

Nearly all of the Pinopolis homes of the early 1800s were built by the same general plan, facing south, two-story, two main rooms with high ceilings on the front downstairs. Each of these two rooms had matching doors [see photo page 3] *opening on a wide front porch, which extended across the full width of the house and around both sides. (The Northampton and Chelsea summer homes also have double doors opening to the piazza.) The rooms across the back extended one room wider than the front, with doors from each extension opening on the side porch. There was no central hall from front to back. The two front rooms were used as "Parlour" and dining room. The enclosed stairs landed in a central back hall, with a door at the foot. Kitchens were never inside the house. They were reached from a back porch, or from a long, covered walkway. Each home had a building, not attached, called "The Lodge." Generally two rooms and a dressing room. This housed the young men of the family and the numerous cousins and friends who fled the heat of Charleston in summer. The lodge also served as a schoolroom with a private tutor for the children in winter.*

that gave dignity and conserving tone to the St. John's Hunting Club of a half-century ago, and which loom on the horizon of memory into a magnitude which is perhaps nearer their true proportions than that which appealed with the diminishing effect of familiarity to the contemporary eye, is that of Mr. William Cain, of Somerset. Tall of stature, dignified in presence and deliberate in all his movements, Mr. Cain exhibited to all a gentle courtesy and polished address which testified conclusively that these traits were not the exclusive heritage of pure Huguenot descent . . . like the others, Mr. Cain was a successful planter of long cotton and his crops usually brought the top of the market. I do not know any better illustration of his courteous good humor than the pleasant and unprotesting smile with which he accepted the statement of a gentleman who was credited with seeing many things and telling more, that walking through the streets of Paris, he had been surprised and gratified at seeing a number of familiar round bales of St. John's Berkeley, all marked W.C. [William Cain] (Wilson 1907).

The children of William Cain and Anne Palmer were William Henry Cain (1834–1909); John Calhoun Cain (1835–1916); Dr. Joseph Palmer Cain (1836–1903); Anna Maria Cain (1841–1905); Edward Greenland Cain (b. 1845); and Julia Anna Cain (1831–1836). In 1886, after her Cedar Spring summer house in Pinopolis burned, Mrs. Kate Porcher lived at Somerset Plantation with her friend, Anna Marie Cain (Caroline Cain 1958).

William Cain left all of his property to his children. Son Joseph and his family continued to live at Somerset for many years after his father's death. Dr. Joseph Palmer Cain and his wife, Mary Catherine Macbeth Cain (1851–1926), raised a large family of twelve children at Somerset Plantation. When Joseph Cain died in 1903, the other heirs of William Cain sold their interests to Dr. Cain's widow, Mary, and her son, William. John Henry Porcher noted in his diary: *In May 1904, the family of the late Dr. Joseph Palmer Cain (1836–1903) purchased Somerset at public auction for $19,000*

This is the Somerset Plantation house as it appeared in 1939. Benjamin Mazyck built the home; William Cain remodeled it in 1840. *Courtesy of the Cain files, South Carolina Historical Society.*

(J. Porcher). Some present family members think the purpose of the auction was to establish the value of the property.

Somerset Plantation continued to be the home place for Cain aunts, uncles, and cousins. Dr. Joseph Palmer Cain's son William (b. 1874) married Elizabeth Ravenel Lucas in 1907. They raised their children at Somerset: William Cain (1908–1993) of Darlington, Charles Lucas Cain (1910–1981), and twins, Joseph Palmer Cain (b. 1914) and Elizabeth Ravenel Cain Boykin (1914–1976).

Somerset plantation was still owned by the Cain family in 1939 before Santee Cooper acquired it by condemnation. The Somerset Plantation house was dismantled before the area was flooded.

The Somerset Plantation summer house as seen in 2005. It was built by William Cain about 1844. In 1985, Mark and Jan Dennis completed extensive interior modernization. This home still has the appearance and feel of its original condition.

Inset: Shown as it is today, this building was the Somerset Plantation summer house office.

Somerset Plantation Summer House

2006 Pinopolis Road (map: 32)

It is thought that William Cain built the Somerset summer house around 1844. It looks much the same today as it did one hundred years ago. The Somerset summer house has been modernized on the interior for comfort but overall still remains quite rustic. I remember very well the long joggling board that sat on the piazza of the Somerset summer house. The joggling board is now in Cheraw at the home of Sarah Cain Spruill, great-great-granddaughter of William Cain.

During my childhood, Caroline "Carrie" Cain (1883–1958), Katherine "Kitty" Cain (1875–1955), and Malcolm Cain (1888–1970), all children of Dr. Joseph P. Cain (1836–1903), lived in the summer house.

Somerset summer house had a lodge that served as sleeping quarters for the boys of the Cain family. Over the years, the lodge has been rented to a succession of people: the Edgar Davis family, Sarah Schuman, and the Joseph Earle Simpson family. Simpson worked at Santee Cooper. His wife, Nannie Louise "Polly," was a nurse at the Berkeley County Hospital. Their son, Joe, who is a friend of mine, now lives in Penn Yan, New York. Joe remembers that *the house* [Somerset summer house lodge] *was very small. It had one bedroom, one bath, a living room/kitchen, and a porch. I slept on the porch. The house was rough with wide pine floors* (interview by author, April 2006).

There was also a small cottage to the east of the Somerset summer house, which Henry R. Dwight Jr. remembers as being a servant's quarters. Marianne Davis Monroe, daughter of Edgar Davis, lived in the cottage when her family first came to Pinopolis. She remembers that the cottage was very primitive. Kinloch Bull, his wife Julia, and adult son, Elias Ball Bull, who was to become one

This 1970 image shows a structure that was known as "the old kitchen" for the Somerset summer house. In the first printing, it was incorrectly stated that this structure no longer exists. Jan and Mark Dennis gave it to Robert Norris in 1990 after it was severely damaged by Hurricane Hugo. Robert moved it to St. Julien Plantation in Eutawville, where he restored the woodwork and installed a new tin roof. *Courtesy of the Elias Bull Collection, Charleston County Library.*

of Berkeley County's most prominent historians and preservationists, rented the cottage about 1950. Robert Stevens, a Pinopolis resident of this time period, remembers that the cottage had a small living room and one bedroom. There was electricity but no indoor plumbing. Water was obtained from a hand pump on the back porch. There was an outhouse on the rear of the lot. The cottage burned while the Bulls were living there. After the fire, they moved to a small building on the Wampee summer house property.

The present owners of the Somerset summer house are Cecil and Stephen Spivey.

Somerton Plantation — no longer existing

St. Juliens may have owned the property forming Somerton plantation before 1725. It came into the Ravenel family through marriage. When Daniel Ravenel (1762–1807) of Wantoot died, he left Somerton Plantation to his son, Daniel (1789–1873). The second Daniel stayed a few years at Somerton Plantation before moving permanently to Charleston. He sold Somerton Plantation to Samuel J. Palmer, who in turn sold it to Frederick A. Porcher (1809–1888) in 1831. Frederick Porcher married Rebecca Branford Rhodes in 1832. She died in 1834. Porcher built the Somerton plantation house and moved there in January 1837, intending to set himself up as a planter. When Porcher purchased Somerton Plantation in 1831, it consisted of 962 acres. While in his possession, Somerton was enlarged to 2,025 acres. Although Porcher had attempted to become a planter, he gave up this life in 1848 to become a professor at the College of Charleston. Porcher arranged for kinsmen to oversee the property in his absence.

In 1858, Porcher sold Somerton to William Cain (1792–1878), owner of Somerset Plantation. In 1868, Cain's son, William Henry Cain, was living at Somerton Plantation (F. Porcher 1921). William Cain's will of 1878 stated that his children inherit his property.

The Somerton Plantation house burned in the 1930s, allegedly at the hands of vandals or thieves who stole the woodwork from the house. The land of Somerton Plantation remained in the Cain family until Santee Cooper acquired it through condemnation in 1939.

The Somerset Plantation lodge was built as a dormitory for boys in the Cain family and as a bedroom for males who were visiting the family.

William Henry Cain House (map: W) — no longer existing

In 1893, Charles Macbeth deeded thirty-nine acres (in the area that later became the Walsh airstrip) to his daughter, Henrietta Macbeth Cain (1837–1905),

Frederick A. Porcher built the Somerton Plantation house in 1837. Porcher sold it to William Cain in 1858, and the property remained in the Cain family until the advent of Lake Moultrie. *Courtesy of Richard D. Porcher, PhD.*

wife of William Henry Cain (1843–1909). Maps spanning the years 1895–1911, which were drawn by Emma Gribbin, Maria Ravenel, and her brother, Dr. Jervey Ravenel, all show the home site of William Henry Cain near the point where present-day Lucas Street joins Walsh Drive. Henry Dwight recalls *a house north of the White Hall summer house* [map:Ⓤ] *where Cains lived. This house, which was the William Henry Cain house, later burned* (interview by author, July 5, 2005).

In February 1906, Richard G. Causey, Berkeley County sheriff, rented the Cain house (J. Porcher). Ten years later, Causey and his family bought the Rene Ravenel Jr. house.

I remember that during my childhood, on my father's property, was an ancient dilapidated one-room building with the remains of a chimney. It was located near the site that the three old maps place the William Henry Cain house. It was probably a servant's house associated with the Cain home. The crumbling ruin stood within a large pecan grove on the west side of the airstrip and north of the Langley house (map: 33).

The William Henry Cain house as it appeared in 1895. *Courtesy of the Macbeth file, South Carolina Historical Society.*

The Treasure House
919 Academy Lane (map: 19)

Any house called "The Treasure House" will always attract attention. This small building would likely have never appeared in Anne S. Ball's book about Pinopolis had it not been called "The Treasure House." Ball's history about "The Treasure House" leaves one wanting more information. Ball simply stated: *Miss Kate Porcher lived with her friend, Anna Maria Cain* [1841–1905], *at Somerset for some time. After the latter's death, Miss Kate left this part of the country. She* [Kate] *built the small house known as "The Treasure House," in which to store her possessions. She was particularly fond of Caroline Cain* [Caroline Porcher Cain, 1883–1958] *and left the house and lot near the old Rectory to her. It was seldom used and so had the fascination of an intriguing mystery* (Ball 1983).

Ball's story of "The Treasure House" raises one's curiosity. Who was Kate Porcher, why did she live with Anna Maria Cain, and what did Kate store in "The Treasure House"?

It turns out that Miss Kate Porcher was not a "miss" after all. She was Mrs. John Palmer Porcher Jr. Catherine Cordes Deveaux Porcher, Kate Porcher as we know her, was the daughter of Stephen Gough Deveaux (1777–1860) and his second wife, Anne Blair Snowden. Kate Deveaux married John Palmer Porcher Jr. (1836–1864), owner of Cedar Spring Plantation, on April 29, 1862. Kate and John lived at Cedar Spring and at their summer house in Pinopolis. A portion of this property was given to Trinity Episcopal Church to be used as the site of its rectory.

John Porcher Jr. died in 1864. His will stated that Cedar Spring Plantation and his home in Pinopolis be left to his wife, Catherine, and his son, George (Bull, South Carolina Historical Society).

So, why would Kate Porcher, who owns Cedar Spring Plantation and a Pinopolis summer house, need to move to Somerset to live with Anna Maria Cain?

Kate Porcher built the Treasure House in 1886 to store her belongings after her Pinopolis summer home burned. This image was made in 1970. *Courtesy of the Elias Bull Collection, Charleston County Library.*

Thomas P. Ravenel noted in his journal on December 1886:

1st: Earthquake rumbling

2nd: Shake about 8 A.M.

4th: Shake about 6 A.M.

7th: Fires general about the country. Very dry & high wind. Fire in woods came down in Pinopolis. Mrs. Kate Porcher's house was burned, Calhoun Cain's house was on fire & was saved by miraculous efforts of those present. There was great danger that the whole village would be burned as many wells gave out. The smoke was so dense that men would almost ride over persons before seeing them & when Mrs. P's [Kate Porcher] *house was burned, those a hundred yards away did not see the fire. Those who were there describe it as a day of horror & say they would rather go through another earthquake than such a reign of fire* (T. Ravenel).

The Treasure House as it appeared in 2005. It is presently located on the site of the Pinopolis Academy.

Having no place to live in Pinopolis after the Cedar Spring Plantation summer house had burned, Kate Porcher moved to the Somerset Plantation summer house to live with her friend, Anna Maria Cain. Anna Maria was living with her brother Dr. Joseph P. Cain and his large family. Caroline "Carrie" Cain, daughter of Dr. Joseph P. Cain, said in her unpublished history of Somerset Plantation that after William Cain died, his son, Dr. Joseph Palmer Cain, lived at Somerset Plantation with his family, his sister Anna Maria Cain, and a friend (Caroline Cain 1958). That friend would be Kate Porcher. It is likely that in summer, Kate Porcher would have lived at the Somerset Plantation summer house.

Moreover, it is also likely that "The Treasure House" was not built because Kate was moving away from Pinopolis and needed to store her possessions but because Kate needed a place to store what she might have salvaged from the fire that burned her home. "The Treasure House" was either spared from the fire or was built afterwards by Kate to store her possessions while she lived with Anna Maria Cain at Somerset.

Anna Maria Cain died in 1905. Kate Porcher then moved away. Where did she go? In the 1920 U.S. Federal Census, Kate Porcher was recorded living in Charleston with Clelia Missroon and others, who were relatives of Frederick A. Porcher, the uncle of Kate Porcher's late husband, John P. Porcher Jr.

When Kate Porcher died in 1923, she left "The Treasure House" to Caroline "Carrie" Cain. About 1930, Cain put "The Treasure House" up for rent. "The Treasure House" had no running water, electricity, or heat; however, Cain had a well drilled for a hand water pump. The rent was five dollars per month.

I remember "The Treasure House" in the 1940s as being the home of Bill and Dora Wiley. The little house was about one hundred yards southwest of the Episcopal Rectory where I lived at the time. "The Treasure House" sat to the right of a dirt road running from Jacob's Lane to the Pinopolis Academy and running parallel to Pinopolis Road. This dirt road no longer exists. When Hal and Elliott West owned the old Kinloch Bull home, which had been built around the Pinopolis Academy, West purchased "The Treasure House" and had it moved to the southern end of his lot. Steve Buckler and his wife, Susan Hillis, who are the current owners of the Pinopolis Academy building, now use "The Treasure House" as a utility building.

White Hall Plantation — no longer existing

White Hall Plantation, located about five miles northwest of Pinopolis, in Colonial days was once the site of a tavern and inn on the Congaree Road. F. M. Kirk suggests that White Hall was named for Blake Leay White, who was commissioner for high roads in St. John's in 1748. Blake White's son, John Blake White (1781–1859), was a distinguished artist and had several paintings hanging in the U.S. Senate lobby in Washington, D.C.

Col. Thomas Porcher (1766–1836) of Ophir purchased the White Hall Plantation property. Colonel Porcher built the White Hall Plantation house about 1824 for his son, Thomas Porcher (1796–1843), who married Catherine Gaillard (1795–1874), daughter of Peter Gaillard (1757–1833) of the Rocks Plantation. White Hall Plantation came into the Lucas family through the marriage

Col. Thomas Porcher built the White Hall Plantation house in 1824 for his son, Thomas Porcher. This image was made ca. 1939, just before the building was dismantled. *Courtesy of the South Caroliniana Library.*

of their daughter, Elizabeth Gaillard Porcher (b. 1825), and Dr. Charles Lucas of Charleston. Lewis Simons Lucas (1853–1915), son of Charles and Elizabeth Lucas, married Florence Hartley Ravenel. Elizabeth Ravenel Lucas (1883–1941), daughter of Lewis and Florence Lucas, married William Cain (1874–1926). The William Cains lived at Somerset Plantation. Lewis and Florence Lucas's other children, Henry Ravenel Lucas (1891–1973) and Florence LeNoble Lucas (1889–1977), never married but lived together at White Hall Plantation and at the Pinopolis summer house.

The White Hall Plantation house was dismantled in 1939 to make way for Lake Moultrie. A replica of the White Hall Plantation dining room using the original woodwork is presently displayed at the Museum of Southern Decorative Arts in Old Salem, North Carolina.

Thomas Porcher of White Hall moved the White Hall Plantation summer house from Goshen Plantation to Pinopolis in 1837. This front view photo was taken in 1970. *Courtesy of the Elias Bull Collection, Charleston County Library.*

Inset: White Hall Plantation summer house, rear view, in 1970. *Courtesy of the Elias Bull Collection, Charleston County Library.*

White Hall Plantation Summer House (map: Ⓤ)

In the early 1800s, Thomas Porcher (1796–1843) of White Hall Plantation was planting at four plantations. In May 1837, he decided to move one of his houses located on Goshen Plantation to Pinopolis to serve as a summer dwelling (P. Porcher). The White Hall summer house passed with White Hall Plantation into the Lucas family. Florence and Henry Lucas occupied the summer house during my childhood years in Pinopolis. It was dismantled sometime after Florence Lucas's death in 1977 (interview by Elias Bull).

Bunker Hill Plantation — no longer existing

Bunker Hill Plantation was thought to have been a part of Hogswamp Plantation. William J. Dennis (1802–1854) purchased Hogswamp Plantation in 1836 from Dr. Edmund Ravenel of Charleston. Robert Press Smith later purchased it from Dennis. T. P. Ravenel noted that in 1846, Bunker Hill was owned by Mrs. Smith (Margaret Cantey Smith) (1787–1854). Ravenel recorded in his journal on December 11, 1854: *Henry F. Porcher, Henry L. Stevens, and myself appraised at Bunker Hill the property of the late Mrs. Margaret Smith* [Mrs. Robert Press Smith], *about 16 negros, stock of cattle and household furniture accounting to about $8,000—besides some bank stock, most of the property left to Miss Susan Dubose.*

Harlock Huxford Harvey (1806–1883) acquired Bunker Hill from Miss Susan DuBose about 1855. While living at Bunker Hill Plantation during the Civil War, Harvey and his family survived two Yankee attempts at burning the house.

The property pictured on page 98 remained in the Harvey family until acquired by Santee Cooper for Lake Moultrie.

The Bunker Hill Plantation house was built in or before 1836. It remained in the Harvey family from 1855 until 1939. *Courtesy of Santee Cooper.*

Harlock Huxford Harvey Pinopolis House
(map:Ⓨ)

Harlock Huxford Harvey (1806–1883) was a son of John Harvey (1778–1823) and a grandson of Arnold Harvey (1754–1834), from whom all Pinopolis Harveys are descended. Sometime after 1855, Harlock Huxford Harvey built a summer house on Broughton Hall Road in Pinopolis on a lot purchased from Charles Macbeth of Wampee. Harvey attached a structure to his house by way of a porch. From this addition, he operated a mercantile store. He continued farming at Bunker Hill but lived with his family in the Pinopolis home on Broughton Hall Road.

Harvey married Margaret "Eva" Evans (1825–1904), and they had three children: Joseph Arnold (1851–1937), associated with the Joseph Arnold Harvey House; Catherine Almira (1853–1922), known as Kate, whose story is in the Kate Harvey House; and Harlock Webb Harvey (1867–1930), associated with the H. W. Harvey House and Store.

Harvey's daughter, Kate, married her first cousin, Thomas Middleton Harvey, in 1878. Kate and Thomas lived with her parents at the Broughton Hall Road house until Thomas became Berkeley County auditor. At that point, they moved to Mount Pleasant, the county seat.

Harlock Huxford Harvey died in 1883. Soon after that, his nephew, John Purkey Harvey, and wife Ann moved into the Broughton Hall Road home with Harvey's widow, Margaret. John P. Harvey operated the old store until about 1890 when he built his own store nearby.

In 1892, when Thomas Middleton Harvey died, his widow, Kate, moved back to Pinopolis with her children and lived with her mother, Mrs. Margaret

Harvey, in the Broughton Hall Road home. In 1895, Kate Harvey purchased the Hogswamp Plantation summer house but continued living with her mother on Broughton Hall Road until her mother's death in 1904.

In 1904, when Margaret Harvey died, her son Joseph bought the shares of the Broughton Hall Road property from his sister, Kate, and brother, Harlock. Kate Harvey then moved to the Pinopolis home she had previously purchased, and it became known as the Kate Harvey House.

Other families who lived in the Harlock Huxford Harvey house were Herman Rudloff and his wife, Julia Elizabeth McCants, along with Julia's parents. George Riggs and his family later owned the house but sold it to Grover Cleveland DeHay in 1945. DeHay lived there but later dismantled the house and used the material to build a smaller house on the lot. His widow, Hazel DeHay, sold the lot to Lee and Jo Anne Brooks in 1967. Grover Cleveland DeHay's smaller house has since been dismantled.

Harlock Huxford Harvey built this house around 1855. It was dismantled in 1954. Left, Joseph Harlock Harvey; center, J. F. Brooks; right, George Riggs. *Courtesy of the J. Brooks Collection, Berkeley Museum.*

Joseph Arnold Harvey House

737 Bunker Hill Lane (map: 4)

In 1882, Joseph Arnold Harvey (1851–1937) married Eva Eugenia Bobo (1854–1905) of Glenn Springs. They moved from Bunker Hill to Pinopolis in 1887. He purchased 35.75 acres of land at the present home site and built his home of timbers from a house he had dismantled.

Joseph Arnold Harvey, known as Joe, was the son of Harlock Huxford Harvey (1806–1883) and Margaret Evans Harvey (1825–1904); the grandson of John Harvey (1778–1823) and Jemima Huxford Harvey (1784–1841); and the great-grandson of Arnold Harvey (1754–1834) and Cynthia Wood Harvey (1761–1812). Joseph Harvey grew up at Bunker Hill Plantation and at his parents' summer house on Broughton Hall Road. Joe's education began in an informal fashion in his father's store on Broughton Hall Road. Richard Yeadon Macbeth

The Joseph Arnold Harvey House in 2005.

Inset: Joseph Arnold Harvey built this home for his family in 1887. As shown in 1939, it still had its original features. Because of extensive hurricane damage, the home has been completely remodeled. *Courtesy of the J. Brooks Collection, Berkeley Museum.*

of Wampee Plantation stopped by the store every day and gave Joe lessons in arithmetic, spelling, reading, and writing (A. Ball 1983). Joe apparently took well to his studies, becoming a successful farmer and politician. Joseph Arnold Harvey was president of the Berkeley County Farmer's Alliance from 1889 to 1891. He served several terms in the State House of Representatives and served three terms as a state senator. Joe was known as "Fence Post Joe" because while in the legislature, he would wait to see how others voted before voting himself.

Joe and Eva Harvey had a daughter, Julia Eloise Harvey, who died in 1909 at age twenty-five, and a son, Harold Huxford Harvey (1888–1960). When Joe Harvey died in 1937, his son, Harold, inherited Bunker Hill Plantation and the Pinopolis property. Harold farmed Bunker Hill Plantation and also operated the Harvey-Lewis Hardware store in Moncks Corner until 1951 when Homer Dantzler purchased the store. Harold married Annie Lucille McCants, and they had four children. Harold Huxford Harvey died in 1960, leaving his property to his son, Harold Huxford Harvey Jr. (b. 1920). Harold H. Harvey Jr. attended Pinopolis School until the third grade when the school closed. He then completed his education at the county schools in Moncks Corner. Harvey remembers Pinopolis school friends, like Markley Dennis, Davis Hare, and Carleton Massalon, and two of his teachers at the Pinopolis Public School: Miss Harrison, who boarded with the Dennis family, and Mrs. Edwards, mother of Margaret Edwards Stevens. Harold H. Harvey Jr. married Frances Arnold (b. 1924).

Harold and Frances Harvey have four children: Linda, Harold III, Joseph, and Robert. After Harold and Frances Harvey built a new home on the east side of Pinopolis Road, their son Hal (b. 1951) and his wife, Naomi Miller Harvey (b. 1950), moved to his father's house, which is the Joseph Arnold Harvey house.

The house suffered severe storm damage during Hurricane Hugo in 1989. Extensive exterior and interior changes have been made to the home.

John Purkey Harvey Home
(map: 35)

John Purkey Harvey Store — no longer existing
121 Broughton Hall Road (map: ⓧ)

John Purkey Harvey (1851–1897), known as J. P., was the youngest son of

Thomas Joseph Harvey (1815–1880) and his wife, Jane Margaret Purkey Harvey (1825–1888). J. P.'s father, Thomas Joseph Harvey, owned Simpson's Plantation, which bordered on the Santee Canal, and operated the lock at Simpson's from 1848 until 1851, when the Santee Canal closed. J. P. Harvey was one of eleven children raised by Thomas and Jane at Simpson's Plantation. Five of their children eventually made their homes in Pinopolis.

In 1882, J. P. married Ann C. Stevens (1856–1892), daughter of Joseph Goodlow Stevens and Jane Chandler Stevens. J. P.'s uncle, Harlock Huxford Harvey, who died in 1883, had operated a store from his home on Broughton Hall Road in Pinopolis for many years. It is likely that after his uncle's death, J. P. and Ann Harvey moved in with his Uncle Harlock's widow, Margaret, and operated the store. Also living there in her mother's home were Kate Harvey and her husband, Thomas Middleton Harvey.

In 1890, J. P. Harvey purchased land across the road from his Aunt Margaret's to build his own home and store from pines cut from the property. It is uncertain how long the John Purkey Harvey store remained in operation. However, on two maps (Marie Ravenel's and Emma Gribbin's) of Pinopolis from the early 1900s, a building identified as "Miss Lou's Store" appears at the end of Broughton Hall Road. Although identified as "Miss Lou's," this would actually be John Purkey Harvey's store, which his sister, Miss Louisa Harvey, operated after his death in 1897.

Ann Harvey died in 1892. J. P.'s unmarried sisters, Sara and Louisa Harvey, moved in with him to help raise his four children and to help out at the store. In Marie Ravenel's memoirs about her childhood in Pinopolis, she states that at the end of the village [Pinopolis], two old ladies ran a shop that sold staples. Very probably, Ravenel is referring to the two sisters, Louisa and Sara Harvey. It is likely that the J. P. Harvey store ended operation about 1910 when Louise Harvey died. In 1980, using a metal detector, I found a dime dated 1910 under an old board in front of the dilapidated store.

Business records from the J. P. Harvey store in Pinopolis give us names of some of the people who were living in the area at that time. An account book belonging to J. P. Harvey for 1892–1893 lists the following Pinopolis residents: R. R.

Inset: John Purkey Harvey built his store adjacent to his house in 1890. After his death in 1897, his sisters ran the store for several years. The store was dismantled in 2004. *Courtesy of Dora Wiley.*

John Purkey Harvey built this home in 1890. It has remained in his family since then.

Lindsay; J. P. Cain; Jake Carson; H. F. Porcher; Rene Ravenel; J. G. Stevens; J. C. Cain; Charles Stevens; W. H. Cain; Mrs. Henrietta Cain; T. P. Ravenel; P. R. Porcher; R. Y. Macbeth; Dr. Dwight; Kate Harvey; and A. J. Jones.

John Purkey Harvey's daughter, Vivian Jane (1885–1954), inherited the home and store. Vivian married William L. Wiley. At her death in 1954, the old J. P. Harvey property passed to Vivian's son, Stevens Harvey Wiley (b. 1913), and then to his son, James Richard Wiley, who is the present owner.

The old J. P. Harvey store building was condemned in 2004 and dismantled. Larry Motes, a Summerville resident with an interest in preserving old buildings, acquired the store's original timbers from his friend Richard Wiley, the present owner. Larry Motes rebuilt the J. P. Harvey store on a smaller scale on his property in Summerville, using the original front doors and the flooring of the old store. When the J. P. Harvey store was dismantled in 2004, it still contained many of the items that had been sold there one hundred years ago. These items are presently displayed in period shelves and cabinets in Larry Motes's reconstructed store in Summerville.

Robert Lindsay House

155 Broughton Hall Road (map: 36)

John Purkey (J. P.) Harvey's sister, Eudora Harvey (1845–1907), married Robert R. Lindsay (1843–1905) in 1875. On land purchased from her brother, Eudora and Robert Lindsay built a small four-room house on the south side of Broughton Hall Road, just east of the J. P. Harvey home. Robert and Eudora Lindsay's three children were born in the small Pinopolis house. Their daughter, Martha Jane (1888–1913), married Charles Alford Ashley Harvey (1862–1931) in 1907. Charles A. Harvey had been married twice before and already had seven children. Charles's children with Martha were Thomas Lindsay Harvey (1908–1971) and Robert Marion Harvey (1913–1970).

In 1917, Joseph Harvey Lindsay (1883–1963), the son of the original homeowners Robert and Eudora Lindsay, purchased the lot on Broughton Hall Road from his nephews, Thomas L. Harvey (1908–1971) and Robert M. Harvey (1913–1970), at a Master-in-Equity sale for $400. Thomas and Robert were very young, ages nine and four.

In 1927, Walter G. Guyton (1856–1929) purchased the four-room home and lot at a sheriff's sale for $15.05. One month later, W. G. Guyton sold the house and property to Louise Guyton Parker, mother of Julia Parker Young, for $10.00 plus other valuable considerations. The valuable considerations were not stated. The house was vacant for many years until 1947 when Louise Parker gave the property to her daughter and son-in-law, Julia and Tom Young. The Youngs enlarged the small house by adding a large room where Julia ran a kindergarten for many years. Julia Parker Young is the present owner of the Robert Lindsay House.

Hogswamp Plantation no longer existing

Hogswamp Plantation was a portion of the estate of Daniel Ravenel (1762–1807) of Wantoot, which passed to his son, Dr. Edmund Ravenel (1797–1870), who built a house at Hogswamp Plantation. Dr. Ravenel was so active in the practice of medicine in Charleston that he never lived at Hogswamp Plantation. In 1836, Dr. Ravenel sold it to his overseer, William J. Dennis (1802–1854), who lived there for eighteen years (F. Porcher 1921, 7–8).

F. A. Porcher lived at Somerton Plantation, which bordered the northeast corner of Hogswamp Plantation. Porcher and William J. Dennis knew each other well. In his memoirs, Porcher recalled an incident in 1846 regarding his neighbor, Dennis:

It is not a pleasant thing to remember the incidents of that summer. There had been a growing dissatisfaction amongst certain people of the parish at the fact that the members of the Legislature should always be selected from the planters, and occasional efforts were made to send some from the other side. This year the division between the two classes [farmer and planter] *was attended with acrimony.*

The head and front of the opposition was a Mr. Dennis [W. J.], *a next door neighbour of mine. He had been an overseer of Dr. Edmund Ravenel, and had been so successful that he had purchased his employer's plantation, and was living upon it with a very respectable force of Negroes. So far as position was concerned, Dennis was a planter* [meaning that if it had been his desire, he could have elevated himself from just a "farmer" to a "planter"], *but as he had made his own fortune, he sought the sympathies of the other class* [the farmer]. *He* [Dennis] *would not be a candidate himself, but exerted himself to set up candidates in opposition to the planter class.* [Dennis offered to Porcher the support of the farmers if Porcher would run for the state senate and be their spokesperson, but Porcher had already given his support to Dr. Peter Palmer, a planter.] . . . *bound by that decision, I quietly declined the offer of Mr. Dennis . . . and though I felt ashamed I could not and would not desert my class* (F. Porcher 1943, 105).

Thomas Porcher Ravenel, a contemporary of Porcher and Dennis, referred to this incident in his journal: *July 4, 1846. Their* [Dennis and the farmers] *first candidate was A. J. Harvey* [1804–1872], *an upright and worthy man, an overseer . . . who gave a short address, rather conciliatory and considerate as there were a few of us of the opposite party* [planter class] *present, he was followed by Wm. J. Dennis, who with a few others prejudiced like himself had brought about all the ill feeling and rivalry in the Parish, he* [Dennis] *made a few hasty and passionate remarks, highly offensive to us, as invited guests, but not worthy of our notice, we left. Oct 14, 1846. They* [farmers] *took their defeat coolly, Harvey being beaten by 16 votes. Dr. Peter Palmer elected Senator, John Harleston and Thos. W. Porcher elected Representatives.*

William J. Dennis did not take the defeat lightly and created a public disturbance. On October 24, Thomas P. Ravenel wrote, *I was summoned by Fred.*

Porcher, Magistrate to give testimony of what occurred on 2nd day of election, as to a riot being created by Wm. Dennis. Ravenel and others were witnesses to the action of Dennis, who along with others, was required to appear before a grand jury on October 26. On October 28, Ravenel recorded: *Appeared before the court sworn and examined* [Dennis and others] *but the grand jury did not bring in a true bill there being some of Dennis's personal friends on the jury, we suppose as we could see no other good reason for throwing it out* (T. Ravenel "Journal").

Seven years later, in 1853, Thomas Porcher Ravenel wrote to his brother Rene, who was in Paris, that William Dennis nearly died during the winter of 1852 and that Dennis was now ill again. Dennis had called for Dr. Morton A. Waring, who refused to see him unless Dennis made up for some offense. Dennis threw himself on the doctor's mercy: *Dr. Waring says he never saw a more humble & penitent man. Dennis is in deplorable condition and can't live very long.* There is no indication from Ravenel's letter as to whether this animosity is related to the 1846 elections.

Doctor Waring's prediction about Dennis's death was accurate, for on May 15, 1854, Ravenel recorded: *I was one of four appraisers of the property of the late Wm. J. Dennis, property consisted of 'Hogswamp' plantation of sixty (60) negroes* (T. Ravenel "Journal").

Six years later, Dennis's property was appraised again: *Feb. 1, 1860, Estate of W.J. Dennis consisting of Hogswamp lot in Pinopolis & 61 negroes was appraised . . . Appraised Hogswamp at $4000, Village lot at $400 & the Negroes at $40,000. One third of the property for his widow* [Sarah Anne McCants Dennis (1816–1888)] *who married R. H. DeHay* [Robert Hamilton DeHay] *& the two-thirds to be divided to his eight children* (T. Ravenel "Journal"). Edward J. and Albert A. Dennis, the second and third sons of William James Dennis, acquired their father's property (Ball 1983). Sarah Dennis DeHay then kept the Hogswamp Plantation summer house in Pinopolis.

In 1879, Thomas Middleton Harvey (1842–1892) purchased Hogswamp Plantation (900 acres) from the heirs of William J. Dennis. Harvey married his first cousin, Catharine "Kate" Almira Harvey (1853–1922), on February 20, 1878. He was the son of Thomas Joseph and Jane Margaret Purkey Harvey and the brother of John Purkey Harvey, owner of the John P. Harvey home and store on Broughton Hall Road. Thomas Middleton Harvey was the auditor for Berkeley County and worked and lived in Mount Pleasant, which was the county seat.

When Thomas Middleton Harvey died in 1892, his wife, Kate, moved from Mount Pleasant to Pinopolis with her children, Lillian Ester (age twelve), Thomas Middleton Harvey Jr. (age eight) and Joseph Harlock Harvey (age five). They lived with Kate's mother, Margaret Evans Harvey, in her home, the Harlock Huxford Harvey house on Broughton Hall Road.

Joseph Harlock Harvey, after graduating from Clemson College, farmed Hogswamp Plantation until Santee Cooper purchased it in 1939.

Kate Harvey House, Hogswamp Plantation Summer House

1905 Pinopolis Road (map: 17)

The appraisal of the estate of William J. Dennis in 1860 included Hogswamp Plantation and a lot in Pinopolis. It is likely that the present Pinopolis home was on this lot at Dennis's death in 1854. In that period, "lot" commonly referred to the plantation summer house. One year after the death of Sarah Dennis DeHay in 1888, J. W. Orvin purchased the Pinopolis property from the heirs of William J. Dennis (Orvin 1950; Ball 1983).

In 1895, Kate Harvey purchased the Hogswamp summer house from J. W. Orvin, and from then on, it was known as the Kate Harvey house. At the time of that purchase, Kate was living with her children and her widowed mother, Margaret, on Broughton Road in the Harlock Huxford Harvey house. Kate Harvey rented out the Hogswamp summer house until 1904, first to Reverend Oliver, a Baptist pastor, and later to a Wiggins family.

Kate's orphaned cousin, Florence McRae Harvey (1868–1946), lived with Kate for many years and was very active in the Pinopolis Methodist Church. When Kate died in 1922, her property was left to her three children. Kate's son Joseph Harlock Harvey (1887–1947) graduated from Clemson, was a surveyor, and farmed Hogswamp Plantation. He served as the Berkeley County farm agent for twenty-five years. Joseph married Evelyn deMedicis in 1929. They lived in the Hogswamp summer house and made some changes, including adding a second story. They had no children.

Hogswamp Plantation summer house was left to Joseph's wife, Evelyn. She later married the Reverend Betts and worked as a Home Demonstration teacher. Avidly opposed to alcohol, she was active in the Women's Christian Temperance Union and became the state president. She came to grammar school classes when I was a boy to speak about the evils and dangers of alcohol. Later, Evelyn and Reverend Betts moved to Columbia. She was killed in 1976 when her car was struck by a train.

Chris and Carmen Biering now own the Kate Harvey house.

The Hogswamp Plantation summer house is also known as the Kate Harvey house. It was built before 1854, possibly by William J. Dennis. This image made in 1918 shows that there was no second story at that time. *Courtesy of Dora Wiley.*

Lewis H. Harvey House

819 Jacob Lane (map: 9)

Lewis Henry Harvey (1869–1924) was the great-great-grandson of Arnold Harvey. Lewis H. Harvey married Minnie Elizabeth Beatson (1868–1929) of Paxville, South Carolina, on March 1, 1894. They lived first in Orangeburg. Then on January 1, 1896, Harvey purchased Crawford Plantation at Longridge near Cooper Store. They had six children while living there. Lewis Harvey

The Kate Harvey house, as seen in 2005, has had several additions since 1918.

Left inset: Kate Harvey at home in Pinopolis in 1918. *Courtesy of Camilla Bishop Henry.*

Right inset: Kate Harvey house living room.

was a farmer and lumberman, and he operated a cotton gin, corn mill, and turpentine distillery.

In 1909, Harvey bought 230 acres in Pinopolis from the William Cain estate. After a daughter died of malaria in 1911, the Harveys decided to move to Pinopolis. They built the existing home in 1912. Until automobiles were available, Harvey traveled the twelve miles to his farm in Longridge by buggy, returning to Pinopolis for the weekends. He was active in the Pinopolis community, serving as chairman of the board of the Pinopolis School for several years. At his death in 1917, the Pinopolis house and Crawford Plantation in Longridge were left to his widow, Minnie Harvey. She lived in the Pinopolis house until her death in 1929. Daughter Marie Harvey lived there until 1951, at which time Marie built a small home across the road from the Lewis H. Harvey house. "Miss Marie" was one of my favorite teachers in grammar school.

Marie Harvey rented the Lewis H. Harvey house to the Henry Kirklands, who lived there for twenty-five years. In 1970, the Lewis H. Harvey house passed to Marie Harvey's sister, Essie Harvey Kizer. Essie Kizer's granddaughter, Anne Marie Lewis Tracy now owns the home.

Lewis Henry Harvey Jr. House

717 Bunker Hill Lane (map: 3)

Lewis Harvey Sr. built the existing home on Bunker Hill Lane for his son, Lewis Harvey Jr. (1906–1968), who married Leona Mae Hewitt (1905–1996) on August 25, 1936 (Harold Harvey interview by author, October 2005). Lewis Harvey Jr. ran an Oldsmobile dealership and oil company in Moncks Corner. He and his wife lived on Bunker Hill Road until Lake Moultrie was filled; then they built a home on the lake near the junction of Pinopolis Road and the Tram Road

Inset: Lewis H. Harvey house dining room

Lewis Henry Harvey built his Pinopolis house in 1912. It has remained in the family since then and is now occupied by his great-granddaughter, Anne Marie Tracy, and her husband, Damon..

Left: Lewis H. Harvey.

Right: Minnie Beatson Harvey.

Lewis H. Harvey built this home in 1936 for his son, Lewis H. Harvey Jr.

(Camilla Bishop Henry interview with author, April 2006). Lewis and Leona Harvey had no children.

The Kunkle family has owned the Lewis H. Harvey Jr. house on Bunker Hill Road since 1992. It has been a rental home for most of its existence.

Harlock Webb Harvey House and Store — no longer existing

Pinopolis Road (map:Ⓑ)

Harlock Webb Harvey (1867–1930), known as Harley, was a son of Harlock Huxford Harvey and his wife, Margaret Evans. Harlock Webb Harvey was the younger brother of Joseph Arnold Harvey, who farmed Bunker Hill Plantation and built the Joseph Arnold Harvey house. His sister, Catherine "Kate" Almira Harvey, is associated with Hogswamp Plantation and the Hogswamp summer house. Harley was educated at Porter Military Academy in Charleston.

In 1891, he married Nettie Camilla Thomas (1869–1915), and they lived at Chicora for several years. Harley was involved in the turpentine business, farmed Bunker Hill and Hogswamp, and had a mercantile business in Chicora. In 1894, Harley purchased the property bounded on the southeast quadrant of Pinopolis Road and the Berkeley Railroad. He dismantled his Chicora home and store and then reconstructed both on the Pinopolis lot as a single building.

Harley served as stationmaster for the Berkeley Railroad and was Berkeley County Master-in-Equity for twenty-four years. Nettie taught at the Pinopolis Academy. She and Harley ran the store. Their daughter, Nettie Elizabeth Harvey (1900–1979), married Ernest Bishop (1897–1972) from Jamestown. Bishop farmed and worked for Santee Cooper.

Camilla Bishop Henry, the daughter of Ernest and Elizabeth Bishop, inherited the H. W. Harvey house and store in 1979. She and her husband, Howard Henry, ran a small antique shop in the old Harvey store for several years. Because the building was deteriorating, they had it dismantled in 1982. Camilla and Howard Henry built a new home adjacent to the site of the old store.

The remnants of the Tram Road can be seen just north of the site of the former H. W. Harvey house and store.

Author's Note: Ernest Bishop's brother, Dennis, was Berkeley County supervisor. Dennis Bishop's wife, Eloise, was office nurse for my father, Dr. Joseph Norman Walsh, for many years. Elliott Williams Bishop, daughter-in-law of Dennis and Eloise Bishop, now lives in the Percival R. Porcher house. George Bishop Jr., grandson of Dennis and Eloise Bishop, now lives with his wife, Jenny, in the Langley house.

Edward O. Hall House

1850 Pinopolis Road (map: 13)

Edward O. Hall was clerk of court and Berkeley County treasurer in 1895 when the county seat was moved from Mount Pleasant to Moncks Corner. Hall purchased a lot in Pinopolis and built a home of black cypress. He and his family moved to Pinopolis on November 16, 1895 (T. Ravenel "Journal"). In 1903, the home was sold to Annie M. Guyton Skinner (1868–1950), the widow of James Bloodworth Skinner (1851–1900), from Georgetown, South Carolina. She had four children: Moselle Skinner (1895–1994), who taught art at Dreher High School in Columbia; James B. Skinner Jr. (1897–1939), an engineer; Harriet Skinner Vardell (1899–1999), a social worker in Moncks Corner; and Julia Skinner (1901–1993), a college librarian.

The original part of the E. O. Hall house remains very little changed. While Annie was living in the house, a two-room wing was added on the east side of the house. The kitchen, which was a separate building on the north side of the house, was moved and attached to the dining room.

At Annie's death in 1950, the house passed to her daughter Harriet Skinner Vardell. Harriet was one of the kindest women I have ever known, and village children were always welcome in her home. When Harriet's brother James died in 1939, his widow, Lillian, and their four children moved into the east wing of the E. O. Hall house and lived with Harriet for many years. The four Skinner children, Mary Anne, James B. Skinner III, William Skinner, and Thomas Skinner, were childhood friends and playmates of mine. Harriet Vardell was a Berkeley County social worker and active in Trinity Episcopal Church. Harriet's son, Dr. James Vardell (1922–1990), grew up in Pinopolis and was a prominent internist in Columbia, South Carolina.

Harriet lived in the Edward O. Hall house from 1939 to 1981, at which time it was sold to John H. Tiencken Jr. He sold the house to Danny and Harriet Dye, who are the current owners.

Harlock Webb Harvey built this home and store in the 1890s from the materials of a home he owned while living at Chicora. This image was made in 1970. *Courtesy of the Elias Bull Collection, Charleston County Library.*

Inset: Pictured is the family of Harlock Webb Harvey. Front row, right to left, are Harlock Webb Harvey; his daughter, Elizabeth Harvey Bishop; and his brother, Joseph Arnold Harvey. Back row left: Annie Lucille McCants Harvey, daughter-in-law of Joseph A. Harvey. Others are not identified. *Courtesy of Camila Bishop Henry.*

Edward O. Hall, Berkeley County clerk of court, built this house in 1895.

Cuthbert Hare House

1842 Pinopolis Road (map: 12)

Sometime before 1901, Harriet "Hattie" Chisolm Cain Wilson (1869–1950) built a home just south of the E. O. Hall lot. Harriet Wilson was the daughter of J. Calhoun Cain (1835–1916) and Charlotte Horlbeck Cain. In 1897, Harriet Cain married Dr. Robert Wilson (1867–1946), who became dean of the Medical College of South Carolina in Charleston. Wilson's father was the Rev. Robert Wilson, MD (1838–1924), who was a surgeon during the Civil War. After the war, Dr. Wilson became an Episcopal priest and never practiced medicine again. John Henry Porcher recorded in his diary that Rev. Dr. Wilson frequently held services at Trinity Episcopal Church in Pinopolis during times when they had no permanent rector.

John H. Porcher wrote that Hattie Cain Wilson often visited in Pinopolis but stayed more with other people in the village than in her own Pinopolis home: *August 7, 1905. Hattie Wilson and children spending a month at Porcher Stoney's.* Village homeowners frequently rented out their homes and stayed with friends or cousins when visiting Pinopolis (J. Porcher).

J. B. Morrison, sheriff of Berkeley County, was renting Hattie Wilson's home in December 1901. In July 1905, he retired and returned to McClellanville. Sheriff Morrison's name was shown on three old hand-drawn maps of Pinopolis. J. H. Porcher's diary cleared up this confusion by explaining that Sheriff Morrison was a renter not a homeowner.

In August 1907, Harriet Cain Wilson sold her house to a county official, Arthur F. Guyton (1876–1954). His wife, Margaret Hare Guyton (1882–1973), known as "Maggie," was a sister of James Cuthbert Hare (1879–1936), who married Joanna Eunice Bolen (1882–1922) in 1911. They bought the former Harriet Wilson property in 1914. According to the 1920 U.S. Federal Census, James Cuthbert Hare was a blacksmith with a wife and four children. Hare's wife, Joanna, died in 1922; in 1929, Hare married Bertha Dennis (1883–1972) from Macbeth.

Hare added a store to the front of the Harriet C. Wilson house. The house and store burned in 1925, but both were rebuilt. The new store was located to the south of the old store and was not connected to the new house, which was a duplicate of the old Harriet C. Wilson house. During the rebuilding, the Hares lived in the Episcopal Rectory. After the home known as the Harriet C. Wilson home burned and was rebuilt by Cuthbert Hare, it became known as the Cuthbert Hare house.

Harold Harvey (b. 1920) remembers being in Cuthbert Hare's store. He says that, in those days, it was customary for a person to hand the clerk a grocery list and wait while the clerk picked out the items. Cuthbert Hare liked to play checkers so much that he sometimes asked the customer to just pick out what he wanted so that he could continue playing checkers.

Harriet Chisolm Cain Wilson, daughter of J. Calhoun Cain, built the first house on this lot around 1897. It burned in 1925 while owned by Cuthbert Hare, who built the present house on the same site soon after the fire.

Hare also built a gristmill to the south of the store. Harold Harvey remembers that Willie Gethers, who also assisted in the store, operated it.

Hare had a blacksmith shop in the rear of the original store, which burned in 1925. When he rebuilt the store, the automobile was dominant, so instead of resuming the blacksmith business, he added gas pumps out front.

Cuthbert Hare's son, Adam Davis Hare (1917–1976), inherited the Cuthbert Hare house and store. Hare's store was the Pinopolis center for scrap metal and tin collection during World War II. Civilians were asked to contribute to the military effort by bringing metal to the store. My friends and I would scour the village for tin cans and any other metal we could find. Once, I carried in a large sack of tin cans and was paid seven cents. With my money, I purchased more candy than I could eat in a day. I can still picture myself as a boy eating those Baby Ruth candy bars and handfuls of Mary Jane candies.

When Davis Hare married his first cousin, Dorothy "Dot" Bolen Hare (1921–2002), in 1948, they converted the store into their home. During this time, the Cuthbert Hare house was rented out. Adam Davis "Davey" Hare Jr., son of Davis and Dot Hare, was born in 1949. His brother, Daniel James "Dan" Hare, was born in 1952. The old store building was torn down in 1971.

Davey Hare and his wife, Sarah, married in 1972. They have lived in the Cuthbert Hare house since making it their home in 1976.

John T. Grant House

1826 Pinopolis Road (map: 7)

John T. Grant purchased a lot from the William Cain estate in March 1902 and built a home there in 1903 (Ball). Charles Cain states that John Tyler, a New England City businessman, built the house for Grant before building his own house next door (Charles Cain, interview by Elias Bull, October 29, 1973, Bull Papers, South Carolina Historical Society).

John T. Grant was manager of affairs for New England City and its depot agent. John's brother, Banks Grant, has been mentioned in some records as being the engineer for the Berkeley Railroad. The Grant house was called the Banks Grant house in a report done for the Historic Preservation Inventory, Berkeley County. The report reads: *The Banks Grant house was built about 1900 by John Tyler and sold to the engineer of the Berkeley Railroad, Banks Grant, who was also manager of affairs of New England City.* Berkeley County records clearly state that John T. Grant was the only Grant owner of the property.

Sometime after his home had been built, John T. Grant removed decorative glass panels from the Berkeley Railroad's only passenger car and installed them in the double swinging front doors of his Pinopolis home. The passenger car, which ran on the Berkeley Railroad from Moncks Corner to New England City, had been purchased from the Savannah Railroad Company. The Savannah Railroad Company had, in turn, purchased the passenger car from a New York cable car company. The Berkeley Railroad passenger car had once been used as a Brooklyn Bridge cable car. The glass panel doors, which were removed from the passenger car, have a detailed border of stylized flowers, which enclose an owl-like figure. The artwork is very unusual, but research has not revealed the design name or the designer.

This image shows a portion of one of the windows from the passenger car of the Berkeley Railroad. Two are now in the front doors of the John T. Grant house.

For most homes in this book, details of property transactions are not given. However, the following transactions for the John T. Grant house are listed as an example of the complexity of family relationships and property ownership in Pinopolis. Involved here are the Guyton, Skinner, Hare, and Parker families. See appendix C for a partial genealogy of these families under Guyton.

Grant sold the home on Pinopolis Road to Minnie E. Wiggins on September 16, 1916. Five days later, it was sold to Margaret Francis Hare Guyton, wife of Rev. Arthur F. Guyton (1876–1954) and sister to J. Cuthbert Hare. Other transactions were: 1919, Margaret F. Guyton to J. W. Mims; 1919, J. W. Mims to Sidney D. Guyton (1873–1953); 1922, Sidney D. Guyton to his brother George N. Guyton (1866–1927); 1927, George N. Guyton to his sister, Annie M. Skinner (1868–1950); 1940, Annie M. Skinner to her grandnieces E. Elizabeth Parker and Julia Parker; 1953, Elizabeth and Julia Parker to their brother, Samuel Pledger Parker; 2000, from Camilla "Millie" B. Parker, widow of Samuel P. Parker to their children, Samuel P. Parker Jr. and Suzanne Parker Hill; 2000, Samuel P. Parker Jr. to Hugh M. and Maxine E. Smith.

John T. Grant, manager of affairs for Chicora, built this house in 1902. He took etched-glass windows from the passenger car of the Berkeley Railroad and placed them in the two swinging front doors.

Descendents of these families still live in Pinopolis. Samuel Pledger Parker Sr. was the stationmaster at the railroad stop in Oakley and also ran a store there. He died from a gallbladder infection, leaving his wife, Louise Guyton Parker, and children, Anne Louise "Big Too"; Samuel Pledger; Elizabeth "Little Too"; and Julia. Louise Parker and the children soon moved to Pinopolis and lived for a while behind Trinity Episcopal Church in a house owned by Connie Murray, who was Lewis H. Harvey's daughter. Louise Guyton Parker and her family later moved to the E. O. Hall house and lived there for about ten years.

At an interview in September 2005, Julia Parker Young told me some of her memories of Pinopolis. Born October 19, 1916, Julia Parker first attended the Pinopolis School and then went to school in Moncks Corner through the ninth grade. Julia's mother, Louise Guyton Parker, wanted her daughter to get the best education possible, so she sent Julia to the Memminger School in Charleston. Julia boarded, along with friends Grace Smith (Mrs. Markley Dennis) and Ruth Smith (Mrs. David Page), in Charleston with an aunt and uncle. Julia Parker went from Memminger to Lander College, where she graduated, excelling in chemistry and tennis. She learned to play tennis on the old Ophir summer house tennis court. Neighbors Anne Sinkler and William Kershaw Fishburne owned the tennis court and welcomed Pinopolis children to play. Julia's teacher was Emily "Cheeka" Fishburne, daughter of Anne and Kershaw Fishburne, and later, wife of Ben Scott Whaley.

Julia Parker Young now lives in the Robert Lindsay house. Julia's nephew and niece, Samuel P. Parker Jr. and Suzanne Parker Hill, are still residents of Pinopolis, having built or remodeled homes of their own.

Harold Harvey remembered that the John T. Grant house had at one time been rented to Mortimer Harvey (1866–1938) and his family: his wife, Gladys, and their children Myrtle, Alva, Clemmie, Virgil, Glaser, Ashley, William, and John (interview by author, October 20, 2005).

On July 29, 1981, a severe windstorm caused several oaks to fall on the Grant house, which suffered extensive damage. Since then, the house has been renovated, and no sign of the large oak trees remain.

The present owners of the John T. Grant house are Hugh and Mickey Smith.

John Tyler House

1814 Pinopolis Road (map: 6)

Both John Tyler (1857–1921) and the previously mentioned John T. Grant were very active in the development and success, although short-lived, of A. S. Emerson's community, New England City, also called Chicora. Tyler had moved with his young family from Cross, South Carolina, to be the first postmaster at Chicora sometime between 1892 and 1900. He built a store, a mill, and his home in Chicora during the town's heyday. When Chicora residents became ill with malaria, Tyler decided to move his family to more healthful surroundings. He took apart his Chicora home and moved it to Pinopolis *board by board and peg by peg* (A. Ball 1983). He began to reassemble his home on Pinopolis Road in April 1901 (J. Porcher). Tyler and his family lived there until the home was sold to Herbert Hucks, who started publication of the *Berkeley Democrat* in 1918. Hucks was probably the second owner of the Tyler house. At Hucks's death, the Tyler house passed to his daughter, Rosa Lou Hucks Wofford, and her husband, Rufus, who had been living in a newer home east of the Tyler house. Rufus Wofford was the Berkeley Grammar School principal and assistant football coach to Jim Bradley at Berkeley High School.

The present owners of the Tyler house are Charles "Chuck" Wofford, son of Rosa Lou and Rufus Wofford, and his wife, Theresa.

Thomas West Williams House no longer existing

Jacob's Lane (map: J)

In 1902, Thomas West Williams (1857–1922) purchased a one-acre lot from Joseph Cain and J. Calhoun Cain. He built a two-and-one-half-story home, elevated ten feet above ground level, for his wife Caroline "Carrie" Wiggins Williams (1860–1926) and two children, J. Russell Williams (1883–1947) and Maude Williams (1885–1966) (A. Ball). On the 1900 U.S. Federal Census, T. W. Williams's occupation was listed as Berkeley County clerk of court. Williams employed a cook, a chambermaid, a laborer, and a janitor.

John Tyler was the first postmaster of Chicora. He dismantled his Chicora house and rebuilt it in Pinopolis around 1901.

During the early 1900s, Charles Wiggins, brother of Carrie Wiggins Williams, his wife Minnie, and their children were renting a Pinopolis home near T. W. Williams (Gribbin 1968). Charles Wiggins Sr. was 36 years old and the Berkeley County auditor. Living with him were his wife Minnie (age 35), daughters Carrie (18) and Maggie (16), son Charles (14), and daughter Ruby (13). In 1902, son J. Lawton Wiggins was born. Two servants, Annette Major and Hampton Major, lived with and were employed by the Wiggins family.

John Henry Porcher recorded in his diary on February 3, 1901: *Mr. Lawton Wiggins* [brother of Charles Wiggins Sr.], *vice-chancellor of the University of the South at Sewanee, on a visit to Pinopolis.*

Thomas West Williams built this house around 1902. The house burned in 1958. Williams's grandson, J. Russell Williams III, and his wife, Sandra, have rebuilt on the property. Left to right: T. W. Williams, J. Russell Williams Jr., and Carrie Wiggins Williams. *Courtesy of LaClaire Laffitte.*

J. R. Williams Sr. married Olive Exum Bardin (1888–1963), who was an artist and lived much of her married life away from South Carolina. Olive Williams lived her later years in Moncks Corner near her son, J. R. "Buster" Williams Jr.

During my childhood, "Miss Maude" Williams, sister of J. Russell Williams Sr., lived in the T. W. Williams house in Pinopolis and operated a tearoom on Main Street in Moncks Corner. It is not known when she first opened her tearoom; however, in the 1930 U.S. Federal Census, she is listed as a tearoom manager. The tearoom, which had green cedar shake siding and a small front porch, was directly across the street from the Berkeley Grammar School. Miss Maude served lunch, which was usually vegetable soup with an egg and olive sandwich. The tearoom was much later moved to West Street where it remains today.

J. R. Williams Jr. (1909–1973) married Clara Lee Peeples (1912–1992). Their daughters are LaClaire Williams Laffitte and Angela Williams. Their son, J. R. "Rusty" Williams III, and his wife, Sandra, operate the Williams Farm Supply in Moncks Corner. Rusty's grandfaher founded the business.

The T. W. Williams house was destroyed by fire in 1958. Rusty and Sandra Williams have built another home on the property.

Jacob's House

Jacob's Lane (map: 10)

T. W. Williams probably built Jacob's house, a small two-room servant's quarters, at the same time that he built his own house, around 1902. The small house was later occupied by Jacob Bash, who was a houseman, servant, and cook for J. Russell Williams Sr. and later for Miss Maude Williams. The 1910 U.S. Federal Census lists Jacob Bash, aged twenty, as a servant.

Jacob was a colorful character in his starched white uniform and cap as he performed his domestic duties for J. Russell Williams. He lived alone in his two-room house but was known to have a group of poker friends (interview Henry Dwight).

The street adjacent to the Pinopolis Methodist Church, which leads to Jacob's house, is now called Jacob's Lane. After the death of T. W. Williams, his home and property were left to his children, J. R. Williams Sr. and Maude. Today, Jacob's house is unoccupied. It stands northeast of the home of the present owners, Rusty and Sandra Williams.

Samuel Porcher Stoney House — no longer existing

Pinopolis Road (map: Ⓡ)

Jacob Bash was a servant in the Williams household for many years and lived in this cottage, now known as Jacob's house.

Inset: Jacob Bash as he appeared in the 1940s. *Courtesy of Sandra Williams*

This home, originally across the road from the Post Office, was built about 1895. Charles Cain (Bull Papers, October 29, 1973, South Carolina Historical Society) says it was built before 1900 for Samuel Porcher Stoney (1850–1922) and his wife, Eliza Chaplin Croft (1854–1927). Stoney purchased Gippy Plantation from John St. Claire White in 1895 and planted there until 1910. Samuel Porcher Stoney lived in Pinopolis with his extended family: his brother, Isaac Dwight Stoney (b: 1842); his twin sister, Anna Maria Stoney (1850–1927); and their six children: Dr. Randell C. Stoney (1876–1906); Charlotte Stoney (1878–1963); Samuel Porcher Stoney Jr. (1882–1960); Laurence O'Hear Stoney (1884–1906); Thomas Porcher Stoney (1889–1973); and Sanford Barker Stoney (1891–1971). Thomas Porcher Stoney was mayor of Charleston from 1923 to 1931.

In 1927, Nicholas Roosevelt, who in 1915 married Emily Sinkler, sister of Anne Sinkler Fishburne of Pinopolis, purchased Gippy Plantation. S. P. Stoney Jr. managed Gippy Plantation for Roosevelt (Irving 1969).

The Samuel Porcher Stoney house in Pinopolis was rented to several families before its demolition.

Stony Landing Plantation

The site was first called Stone Landing because of the copious amounts of marl found in the area. It was then mistakenly called Stoney Landing before finally being known as Stony Landing. Stony Landing was at the head of Biggin Creek, which led into the upper Cooper River. The landing served as a center of commerce and trade between Indians and the interior. Supplies moved to and from this landing as far west as the Mississippi River by way of the Cherokee Path, Congaree Road, and interior trails. Later, rice planters in Middle St. John's carried their produce by wagon to Stony Landing to be shipped to Charleston on

Samuel Porcher Stoney, owner of Gippy Plantation, built this house about 1895. The image was made in 1970. *Courtesy of the Elias Bull Collection, Charleston County Library.*

the Cooper River. A large warehouse at the landing provided storage until ships arrived. A general store operated at Stony Landing from 1731 to 1751. Stony Landing declined as a trade center around 1800 when the Santee Canal opened and new roads were being built (www.oldsanteecanalpark.org).

Stony Landing Plantation was the last part of the Colleton Fairlawn Barony to be disposed of when 2,319 acres were sold to John H. Dawson in 1839 (Cross 1985). Dawson built the Stony Landing Plantation house around 1843.

During the Civil War, Dr. St. Julien Ravenel (1819–1882) of Charleston owned Stony Landing. Dr. Ravenel designed a semi-submersible submarine called the "Little David" [CSS *David*] that was built on site at Stony Landing Plantation. David C. Ebaugh (1824–1895), who was already working at Stony Landing as manager of the Confederate Nitre Works and Brick Mill, became the chief mechanic and engineer responsible for construction of the submarine.

On October 5, 1863, in Charleston Harbor, the Little David performed the world's first submarine attack against what was the world's most formidable battleship of its day, the *New Ironsides*. The *New Ironsides* did not sink but was heavily damaged (Berkeley County Tricentennial Commemorative Edition, 1970). The Little David was also severely damaged (Irving 1969).

David C. Ebaugh was considered by some to be a local hero for the part he played in the Civil War. However, there is a previously undisclosed side of him. Ebaugh was in the habit of telling people that the Little David had been named for him. Dr. St. Julien Ravenel's wife, Harriott Horry Rutledge Ravenel, contradicted this story by saying that she had named the submarine *David* for the Biblical character David who killed the giant Goliath. Ebaugh also claimed to have designed the Little David. That is also a false claim. Samuel Gaillard Stoney (1853–1926) gave this account of the Little David project: *Lieutenant W. T. Glassell of the Confederate Navy . . . my father, Samuel D. Stoney (1822–1892)*

David C. Ebaugh. *Courtesy of the Department of the Navy, Naval Historical Center.*

and Dr. St. Julien Ravenel were dining together . . . Dr. Ravenel said that he thought he could design and build a spar torpedo boat . . . Ravenel designed and superintended the building of the boat . . . On the occasion of my [Samuel Gaillard Stony's] *visit to Stony Landing, my father did not tell me of where we were going nor anything about the trip . . . this place may have been chosen for its safety and secrecy . . . spies seemed to infest the city* (Irving 1969).

Another account regarding Ebaugh's claim on the Little David and his true sympathies during the Civil War was given by M. C. Orvin: *In 1892 Mr. Ebaugh wrote letters to the Reverend W. H. Campbell, of Charleston, giving an entirely different account of the building of this boat* [the Little David], *claiming for himself all credit for planning, financing, and building it. Let it be borne in mind, however, that in 1873, Mr. Ebaugh filed a claim in Washington for $25,000 "for property taken from me at the evacuation of Charleston," and declaring that his sympathy had always been with the North and that he had been ready at all times to aid the Union cause* (Orvin 1961).

Ebaugh was thought to have profited illegally from his work at Stony Landing during the Civil War. Mrs. T. G. Barker wrote a letter to Charles Lucas of Pinopolis: *Dear Charlie, Charleston, August 6, 1866, Major Rice . . . has in his possession, or expects to have, evidence sufficient to compel Ebaugh to an account of the profits of the Mills before and during the War, and since the War ended. He was disappointed in not finding me in possession of all the information, which Mrs. White and you have about the business and does not want to begin proceedings against Ebaugh until he is quite ready with testimony and until Mrs. White is ready* (Barker, Lucas Papers, South Carolina Historical Society).

Henry William Ravenel in his private journal suspects Ebaugh of certain shady business practices: *Aug. 23, 1865 He* [Rene Ravenel] *gives a deplorable account of the condition of affairs in the country around* [St. John's]. *Robberies are of daily occurrence. Ebaugh, Markley & Easterling & others are buying all the cotton the negroes can steal & carry to them. Everybody is losing cotton, cattle, hogs, and sheep* (H. Ravenel 1947, 272).

David C. Ebaugh's assets at the end of the Civil War were $165,000, an amazing amount when compared to the assets of other Southerners (chenowethsite.com/chenweth.htm). For example, Rene Ravenel, a well-to-do planter, showed assets of $57,000 in 1860 but was reduced to $11,000 in 1870. Ebaugh somehow escaped official investigation and moved to Charleston after the Civil War. While living there, he served as an alderman under Charleston's Mayor Courtney and was a member of the New England Society. Ebaugh is buried at Magnolia Cemetery (*News and Courier*, Charleston, S.C., August 10, 1895).

David Ebaugh had two Pinopolis connections. Ebaugh owned the Pinopolis subdivision property now known as Land-O-Pines, which he sold to Jacob Carson in 1873. He also owned and sold property to A. D. Hare, what is now the A. D. Hare house.

Stony Landing Plantation had a successful lime operation in the 1850s under the direction of Dr. St. Julien Ravenel and Charles Stevens. Ross Hanahan and Samuel Porcher Stoney revived the lime production at Stony Landing in August 1903 (J. Porcher).

E. J. Dennis Jr. (1877–1930) purchased Stony Landing Plantation in 1919. Dennis's father, Gen. E. J. Dennis (1844–1904), was the first in a long line of Dennis men active in law, politics, and government. Referred to as one of Berkeley County's most significant citizens, General Dennis was born at Hogswamp Plantation. He attended Pineville Academy and the Citadel, leaving at age twenty to join the Confederate Army. Dennis had a distinguished military career and was especially known for organizing a company of scouts who were significant in protecting women, children, and the aged left unprotected on plantations scattered throughout the Low Country following the final months of the Civil War. After the Civil War, Dennis became a general in the state militia. He was very active in the early years of Berkeley County politics, and his descendants continue that tradition.

In 1869, General Dennis married Adelaide Markley (1847–1928). They lived on Fair Spring Plantation, which was owned by her father, Wade H. Markley. The children of E. J. and Adelaide Dennis were Walter, William H., E. J. Jr., Louise Dennis Altman, and Julia Dennis Carson. Dennis farmed, had a store, and was a surveyor. He studied law and was admitted to the South Carolina Bar in 1884. He was a state representative and later a senator from 1894 until his death in 1904. Gen. E. J. Dennis is buried at Mount Olivet Baptist Church near Cross.

John H. Dawson built the Stony Landing Plantation house around 1844.

General Dennis's son E. J. Dennis Jr. (1877–1930) attended Clemson College and studied law under his father. In 1906, he married Ella Mae Coney and in 1908 opened a law office in Moncks Corner. E. J. Dennis Jr. farmed while continuing his law practice. He purchased Stony Landing Plantation in 1919. E. J. Dennis Jr. served several terms in the State House of Representatives before becoming a state senator.

On July 4, 1930, Senator Dennis was shot and killed on Main Street in Moncks Corner. The murder was politically motivated and associated with the "moonshine wars" being waged in Berkeley County. W. L. "Sporty" Thornley was tried and convicted of the murder and sentenced to life in prison. After serving eight years and one month in the South Carolina State Penitentiary in Columbia, Thornley was pardoned by Governor Olin D. Johnston and released on December 21, 1938.

Senator Rembert Dennis, son of Senator E. J. Dennis Jr., sold a portion of the Stony Landing Plantation property for what is now the Berkeley Museum and Heritage Center. Santee Cooper purchased another portion of Stony Landing Plantation, including the plantation house, to create the Old Santee Canal State Park.

Stony Landing Summer House
1834 Pinopolis Road (map: 11)

E. J. Dennis Jr. bought the J. Calhoun Cain home and lot (map:Ⓖ) in Pinopolis before 1915 and began spending summers there with his family. Around 1915, Dennis razed the J. Calhoun Cain house and built a new house. Dennis's new house in Pinopolis burned in 1932. The house was rebuilt on the site around 1935. The house built in 1935 is the house still standing today, called the Stony Landing summer house. In 1939, the Fair Spring Plantation house, formerly owned by Gen. E. J. Dennis, was dismantled to make way for the creation of Lake Moultrie. The stairway and two mantels from Fair Spring Plantation were added to the Stony Landing summer house in Pinopolis.

The children of E. J. Dennis Jr. and Ella Mae Coney were E. J. Dennis III (1907–1931); Maxine Dennis Jones (1919–1987); Beatrice "Bea" Dennis Plyler (b. 1912); Sarah Adelaide Dennis Lacey (1913–1987); Senator Rembert C. Dennis (1915–1992); Rupert Markley Dennis (1918–1998); and William "Billy" Albert Dennis, MD (1924–1992), whose name was legally changed to E. J. Dennis after the deaths of his father and brother.

When interviewed at her Greenville home in August of 2005, Bea Dennis Plyler, age ninety-three, was living alone, keeping house, and driving her automobile. She remembers many things about Pinopolis: *We lived in the Calhoun Cain house on the property. It was taken down and a new one built around 1915. I was married in 1932 and sometime after that the house burned. Everything in the house was lost. My mother lived in the garage, which was the servants' quarters. A new house was rebuilt between 1935 and 1938. One day my sister, Maxine, and I decided to walk to Stony Landing. We were picked up by Sheriff Causey in his buggy and taken back home. Mrs. Griffith would come to our home and stay for a week making clothes for me and my sisters. We were supposed to wear hats to school in Moncks Corner. We were told that it was to keep us from getting freckles. No girls from other families wore them, so we would hide ours in Mr. Hare's gristmill and pick them up after school. We went to school in Moncks Corner starting in the eighth grade.*

Cenie, a black woman, would come from the north of Pinopolis in an ox cart, pick up our laundry, take it home and wash it in a big kettle, and

E. J. Dennis built the first Stony Landing Plantation summer house in 1915. It burned in 1932, and the present home was completed in 1935.

return it. We had another servant, Hannah, also called Dah. She took care of the children and kept the pantry keys. I thought she had a royal bearing but she was not aware of it. Once she gave us benne seeds and some chickens and I thanked her. She said, "You don't have to thank me because you are 'Missus.'" Liza, a handywoman, and several men worked around the place. I suppose our financial status was like everyone else—we didn't have much. My father tried lots of things to raise money, even raising sheep and cattle, but was not very successful.

Bea continues: *We had electricity from a Delco battery system* [the Delco house is still on the property] *and we had a well and electric pump. The well wasn't completely covered and I used to worry that Billy, the baby, would fall in. I loved Billy and was the first to hold him after he was born. Mother was a Baptist and went to St. John's. Father went to the Pinopolis Methodist Church and I was married there. Mother didn't approve of dancing. That caused us trouble because dancing was big in Pinopolis. Mrs. Fishburne had dances for children and we wanted to go. Sometimes an ice cream truck would come from Moncks Corner. There was an icehouse that was owned by Mr. Kingbar before Mr. Massalon bought it. Vendors came through Pinopolis selling Stanley products like furniture polish, cleaners, and brushes. Other peddlers would sharpen knives and scissors.*

Every spring we took calomel followed by castor oil and Groves Tasteless Chill Tonic—I think the Grove Park Inn came from that. Dr. E. W. Groves of Paris, Tennessee, built the Grove Park Inn at Asheville, North Carolina, from the millions he made from the tonic. It contained a high percentage of alcohol and was advertised as being good for women and children.

I finished Berkeley in 1927 and graduated from Asbury College in Wilmer, Kentucky, in 1931. I had to borrow money for college and paid it off by teaching school at Cross. Frank Kirk was the principal.

The present owners of the Stony Landing summer house are James and Carla Anderson.

Langley House

239 Walsh Drive (map: 33)

This lot was once part of the thirty-nine acres that Charles Macbeth deeded to his daughter, Henrietta Macbeth Cain (see William Henry Cain House). In July 1900, E. J. Dennis surveyed a one-acre portion of this property for Henrietta Cain (1837–1905). No record was found of Cain selling this lot to B. C. Cook; however, records show that, in 1911, Cook sold the lot to S. E. Law for $75. In the 1910 U.S. Federal Census, Bonner C. Cook, age forty, is listed as a farmer of Second St. John's Berkeley. Living with him were his wife, Rebecca, and four children.

Mr. Law from the Lebanon area of Berkeley County built this house before 1920. It was the home of Berkeley High School teacher E. Jack Langley for many years and became known as the Langley house. This image shows its appearance in 1970. *Courtesy of Jenny Bishop.*

Sometime after 1911, Law built the existing house that is now referred to as the Langley house. A 1920 deed of sale calls

Inset: Dining room of the Langley house.

Langley House in 2005.

this property the Law lot. Law was living in the home now known as the Langley house, but he could not access his lot without going through the Causey property. In August 1920, Law and the Causey family signed an agreement giving Law access to his lot.

S. E. Law died intestate in 1925, and the property passed to his widow, Catherine Law. Catherine died intestate in 1935, and her children became the owners. In 1937, G. W. Law purchased the house and lot from his siblings. The 1930 U.S. Federal Census lists George W. Law, age forty-six, as the manager of a mercantile business.

In 1937, E. Jackson Langley purchased the lot from G. W. Law for $1,400. Henry R. Dwight Jr. (interview, July 5, 2005) related that Jack Langley was a schoolmate in the 1920s. Dwight said that, at that time, Jack Langley's house was known as the Law house. Jack Langley, a longtime Berkeley High School teacher, lived in the Law house from the 1940s through the 1960s. From then on, the Law house became known as the Langley house.

During my childhood, Jack Langley was my next-door neighbor. His invalid mother lived with him until she died. I remember in 1944 that there was a horse barn on the southeastern corner of the lot. It had a hayloft where my friends and I played, and old hay was still present on the floor.

In 1958, Jack Langley sold the lot to my father and his second wife, Sarah Edens Russell Walsh. At the death of my father in 1974, Sarah Walsh became the owner. She sold it to her daughter, Sallie, and husband, Bruce Bonge, in 1981. Sallie and Bruce Bonge sold the property to the present owners, Jenny and George Bishop, in 1993.

Homer E. Dantzler House

1812 Pinopolis Road (map: 5)

Homer E. Dantzler bought the house in 1936 from J. Russell Williams Sr. Dantzler came to Pinopolis from the Calhoun District where he had lived with his parents and farmed. Dantzler, according to the 1930 U.S. Federal Census, boarded with Joseph Arnold Harvey and wife, Eva Eugenia Bobo Harvey, who were Dantzler's uncle and aunt. He worked in his uncle's hardware store in Moncks Corner. Dantzler bought the house just across Pinopolis Road from his uncle and aunt's Pinopolis home on Bunker Hill Lane. Dantzler's sister Inez joined him to share the home.

Homer Dantzler purchased the property and existing home on August 3, 1936, from Spearant Liquidating Corporation, which was owned by J. Russell Williams. Dantzler paid $1,500 for the house and property, which bordered on Lewis H. Harvey and Joseph Arnold Harvey property. The original house has been modified over the years but still retains structural elements. The lot at one time included a stable, a rabbit hutch, several chicken coops, a batten board garage, and a brick springhouse. Dantzler also had a large garden on the property.

Homer and his wife, Novella, raised their daughter, Nola, in their Pinopolis home. Homer died in 1975, leaving the house and property to Novella, who continued living in the home until her death in 1992.

George J. Davis III and his wife, Cecy Guerry, purchased the Dantzler house in 1994. The house has been remodeled somewhat, but bead board ceilings, heart pine floors, batten board exteriors, glass doorknobs, sliding pocket doors, and a millstone sunken into the front walkway reveal the house's age and character.

Russell Williams built this house in the 1930s. It was the home of Homer Dantzler for many years when he operated the Harvey Lewis Hardware Store in Moncks Corner.

Peterson House
(map:Ⓒ)

Emma Gribbin shows this home on her map and mentions in her memoirs that Mr. and Mrs. Peterson are residents of Pinopolis between 1895 and 1911. No information has been found regarding the home or the family.

Massalon House
2033 Pinopolis Road (map: 34)

Henrietta Stevens (1862–1933) married John St. Claire White (1854–1933), owner of Gippy Plantation, in 1881. Henrietta White purchased the land at this location from Henrietta Gourdin Ravenel Macbeth (1816–1888) in 1886. The Pinopolis home may have been moved from Gippy Plantation. Henrietta White's brother, Charles LeNoble Stevens, and his wife, Mary Wharton Sinkler, also summered at their Pinopolis home, the Northampton summer house.

Henrietta White owned the house for only two years and sold it to M. J. Wooten in 1888. In 1892, Wooten sold it to the J. P. Harvey Co. Tallulah Allen Stoney (1851–1935), widow of William Edmund Stoney (1839–1896), purchased the property in 1896. In 1910, Tallulah Stoney sold it for $300 to the St. John's Baptist Church for use as a parsonage. Robert W. Massie Jr. purchased the property in 1921 and, in 1929, sold it to John and Anna Massalon. Their grandson, Thomas Massalon, and his wife, Barbara, now own and occupy it. The interior has been completely remodeled.

Hawkins Jenkins House (map: Ⓜ) no longer existing

Hawkins King Jenkins (1859–1926) was an attorney who practiced law in Mount Pleasant and Moncks Corner. About 1896, he purchased Wantoot Plantation from Richard Y. Macbeth for $1,900 and built a large home in Pinopolis on what is now Morton Waring Lane. Jenkins served in the South Carolina State House from 1888 to 1898 and the senate from 1898 to 1900. He was a lay reader at Trinity Church Pinopolis for fourteen years.

This is the only known photograph of the Hawkins Jenkins house. Sitting on the porch in 1905 from left to right are Henry R. Dwight Sr., Percival R. Porcher II, Hawkins Jenkins Jr. or Manigault Jenkins, Hawkins K. Jenkins Sr., and Isaac de Cherigny Porcher. *Courtesy of Martha Porcher Ravenel.*

His children were Emma Manigault Jenkins (d. 1974), who married the Rev. Robert Emmet Gribbin, an Episcopal bishop; Joseph Evans Jenkins; Hawkins King Jenkins Jr.; and Gabriel Manigault Jenkins (Gribbin 1968).

According to Henry R. Dwight Jr., Henry Severence Scarborough purchased Wantoot and the Pinopolis house from Jenkins. Dwight remembers the following Scarborough children: H. S. Jr., called Hallie; Richard; Eldridge; George; Coke; Bascomb; Allen; Lucille; Mary Alice, who married John Harvey; and Ruby.

Julia Parker Young (b. 1916) remembers, as a teenager, playing the piano in the Jenkins house for a dance and party (interview by author, September 20, 2005).

Around 1930, while owned by the Scarboroughs, the Hawkins K. Jenkins house in Pinopolis burned (Henry Dwight, interview by author, October 20, 2005).

Adam Davis Hare House
324 Pineland Lane off Pinopolis Road (map: 2)

The A. D. Hare house and St. John's Baptist Church, both located near Hyrne's Curve, are not in Pinopolis proper. They are included in this book because of their close association with the old village of Pinopolis.

Solomon Clarke probably built the white, framed, two-story home around 1830. Early land deeds did not give a detailed description of the improvements on the property. An adjoining porch attaches the two-story house to an older one-story structure, presumed built in the late 1700s before Pinopolis was settled. The one-story structure, probably an early store, consisted of two large rooms heated by fireplaces.

The kitchen was a separate building located near the main dwelling. When this kitchen burned, a one-story dwelling was moved and connected to the two-story home to create a new kitchen. In the 1890s, the back porch was enclosed to add bedrooms and more living area.

Early deeds show that Solomon Clarke (d. 1850), J. J. Coleman, and then David C. Ebaugh (1824–1895) in 1861 owned this property. In 1874, Adam Davis Hare, who worked as superintendent for Ebaugh, purchased the house and 2,342 acres from him.

The Adam Davis Hare house was built in the 1830s; A. D. Hare purchased it in 1874. It has been owned and occupied by his family since then.

In 1894, A. D. Hare conveyed the home and property to his two daughters, Annie and Margaret. Annie Hare McCants and her husband, Josiah Owens McCants, purchased her sister Margaret's interest in the property in 1901. The McCants lived there with their eight children. Upon her death in 1965, Annie passed it on to her son Arthur W. McCants Sr. Arthur's son, Edgar McCants, and his wife, Annegell, presently own the home.

Some of the old outbuildings of the Hare property remain, including a carriage house, a smokehouse, a feed barn and a corncrib. Original outbuildings that no longer exist include stables, a cotton gin, and a turpentine mill. There were also numerous small, one-family tenant dwellings scattered throughout the property. These buildings were located near the fields assigned to that tenant family. Only one tenant dwelling, dating back to the early 1800s, remains.

This is the only remaining tenant house on the A. D. Hare property.

Trinity Episcopal Church
Black Oak — no longer existing

Originally, there were ten Church of England parishes in South Carolina. The largest was St. John's Berkeley, which was established in 1706. St. John's Parish was about forty miles in length and was divided into Lower, Middle, and Upper sections. The first church built in St. John's Parish was Biggin Church near Moncks Corner. The church was completed in 1712, burned by a forest fire in 1755, rebuilt in 1756, and burned by the British in 1781. It was rebuilt in 1809 but burned again by a forest fire in 1886. By 1886, it was used infrequently and therefore not rebuilt (Cross 1985). The brick pillars, foundation, and graveyard remain and are accessible to visitors.

Middle St. John's planters attended Biggin Church in the early days; however, about 1800, they separated from Biggin because of their desire to have a church closer to their plantations. At this time, the Rev. P. M. Parker held services on Northampton, the property of Gen. William Moultrie. Northampton was about a mile from an area called Black Oak.

Black Oak got its name from one of the locks on the old Santee Canal. The area around the Black Oak Lock developed into a small commercial center, having a warehouse, store, homes, and the St. John's Hunting Club building. Black Oak was the intersection of several heavily traveled roads and formed a hub for many plantations in Middle St. John's. The Black Oak Agricultural Society met there, and it also became a voting precinct.

Rene Ravenel (1762–1822) gave land near the Black Oak Lock to the Episcopal Church. After the Revolutionary War, the Episcopal Church replaced the Church of England in America because patriots did not want to be identified with the British and King George. In 1808, the first Black Oak Church was built. This building was donated to the Methodists in 1846 (or 1847) and moved to Macbeth. In 1927, the old Black Oak Church building was dismantled and the present Methodist Church constructed on the same site. A new Trinity Black Oak Episcopal Church was built and consecrated by Bishop Gadsden on January 1, 1847.

After the settlement of Pinopolis, some parishioners didn't want to travel back and forth to Black Oak every Sunday and wanted their own chapel in the village. In October 1845, the first service in Trinity Pinopolis was held.

After the Civil War, many planters were forced for economic reasons to leave their plantations. Attendance at Black Oak Church dropped significantly, and it came to be used only occasionally for special services. The Black Oak Church building had to be dismantled in 1941 to make way for the Santee Cooper project, and its parts were sold for $7,000. The Black Oak graves were moved to St. Stephen's Episcopal Church Cemetery and Pooshee Plantation (A. Thomas 1957).

Some of the structural remnants of Black Oak Church are preserved in Pinopolis. Its front and rear doors and flooring are in the old parish house in Trinity Church, Pinopolis. Window shutters from Black Oak Church have been transformed into doors for closets in the entryway of the P. R. Porcher house. The Black Oak pews were moved to Trinity Church, Pinopolis, but were replaced when the new Pinopolis church was completed in 1968. One of Black Oak's pews sits on the front porch of the P. R. Porcher house.

The first Trinity Episcopal Church Black Oak was built in 1808. The second one, shown here, was built in 1846. This image was made in 1939, just before the church was dismantled to accommodate Lake Moultrie. *Courtesy of Santee Cooper.*

Trinity Episcopal Church Pinopolis

1915 Pinopolis Road (map: Q)

In the mid-1800s, many Black Oak Episcopal Church members spent their summers in Pinopolis and wanted to be able to attend services in the village rather than traveling to Black Oak. For this reason, a chapel of ease was built in Pinopolis. A chapel of ease is a building other than the main church of a parish that is more conveniently located for some of the congregation. The minister held services at Black Oak in winter and Trinity in summer. The first service in Pinopolis was held in October 1845.

The Pinopolis congregation struggled financially through the years. At one time, St. Stephen's Episcopal Church and the Rock's Church supported Trinity Pinopolis. Trinity Pinopolis became independent in 1855, and its first rector, Rev. J. J. Sams, served from 1855 to 1860. A rectory was built for him in Pinopolis in 1855. After the Civil War, Trinity Pinopolis could not afford to pay the $600 yearly salary for a priest, so in 1867, Trinity Pinopolis again associated with St. Stephen's Episcopal Church.

A parish house for Trinity Pinopolis was completed in 1944 with funds from the sale of Black Oak Church to Santee Cooper, along with a gift from Mr. Behrmann. As mentioned earlier, the Black Oak floors and front and rear doors were used in the parish house. The front columns of the Pinopolis parish house came from Belvidere Plantation (Ball 1983).

On May 23, 1948, Bishop A. S. Thomas formally consecrated the Trinity Pinopolis Chapel as Trinity Church, Pinopolis. After 122 years of service, the Trinity Pinopolis church building was found to be unsound and restoration was considered impractical. The old church building was dismantled, and a new structure built on the same site. My father, Dr. Joseph Norman Walsh, drew plans for the present new building, which was completed and consecrated in 1968. A new multifunction parish hall was built northeast of the new church building, and the old parish house was renamed Black Oak.

Some of the former rectors were J. J. Sams; P. F. Stevens (see the section on the Nazareth Church); W. O Prentiss, 1876–1886; H. L. Phillips, 1886–1888; James Simonds, 1893; H. T. Gregory, 1895; A. E. Cornish, 1898; John H. Brown, 1900; Dr. Robert Wilson, who performed occasional services; O. T. Porcher and W. S. Holmes, who gave mission services in 1915; A. E. Cornish, who served as rector in 1898 and again in 1915; A. E. Taylor, 1921; Walter Mitchell, 1923; E. B. Guerry, 1932; D. N. Peoples; and my uncle, Tracy Fishburne Walsh, in 1935. Lincoln Taylor, 1937–1946, left Pinopolis to join the Order of the Holy Cross and later returned to Pineville to establish a branch of the Order of the Holy Cross at the home and property of Gen. Francis Marion, then owned by Eljule Marion. Other priests at Trinity were Wallace Martin; Steven Skardon; Harold Barrett; Loren Mead; Jessie Sparks; James Crowther; John Flanigan; David Christiansen; David Simpson; Russell Johnson; Ted McNabb; and Robert Horn. Harold Fenters is the present rector.

This image shows Trinity Episcopal Church Pinopolis in 1967, shortly before it was dismantled because of irreparable deterioration. *Courtesy of Robert Cuthbert.*

Inset: The first service was held here at Trinity Episcopal Church Pinopolis in October 1845. This image shows its appearance in 1900. *Courtesy of the Cain files, South Carolina Historical Society.*

Episcopal Church Rectory in 2005.

Inset: The Episcopal Church Rectory was completed in 1845 for the first rector, Reverend Sams. This image shows its appearance in 1970. *Courtesy of the Elias Bull Collection, Charleston County Library.*

For many years, there were only lay readers serving in Trinity Pinopolis. For much of the twentieth century, it was a mission of the Episcopal Diocese of South Carolina. At present, it is a self-supporting congregation.

Episcopal Rectory

1849 Pinopolis Road (map: 16)

Catherine "Kate" Cordes Porcher, widow of John Palmer Porcher Jr., owner of Cedar Spring Plantation and its Pinopolis summer house (map:(BB)) donated the Episcopal Rectory lot. The Cedar Spring summer house was west of where the rectory now stands.

The rectory was completed in 1855 in time for Trinity's first rector, J. J. Sams, to live there. Sams served from 1855 to 1860. The next rector was Rev. P. F. Stevens, a Citadel graduate, who resigned his church position during the Civil War and entered the Confederate Army. It was Major Stevens who on January 10, 1861, ordered troops on Morris Island to fire upon the Union ship *Star of the West* in Charleston Harbor (*Harper's Weekly*). This act began the Civil War. Because his action at Fort Sumter was known among the Negroes, Stevens's home suffered more damage than anyone else's in Pinopolis. In March 1865, Stevens's home, the Episcopal Rectory, was robbed of everything: furniture, bedding, and clothing. Stevens had to carry his family to a place of safety (T. Ravenel).

In the 1870s, the Episcopal Church refused to allow Negro priests to enter its buildings. In sympathy for the Negro's cause, Stevens left the Episcopal Church to help Episcopal Negroes find a new church alliance. He accomplished this through the Reformed Episcopal Church and later became one of its bishops.

For most of its early history, Trinity Pinopolis was not able to afford a resident rector, so the church rented out the rectory. Anne S. Ball lists several persons or families who were renters: Georgie Palmer Townsend; DuBose Heyward; the Herman Robertson family; Miss Maggie Mae Williams; the Cuthbert Hare family; Mr. and Mrs. James Russell Williams Sr. and their son, J. R. Williams Jr.

Anne Allston Porcher (1882–1970), daughter of Dr. Francis Peyre Porcher, recorded in her diary that she lived in the rectory with her family as a child. Anne said that, in 1898 when she was sixteen, she visited Mrs. T. W. Williams, who was renting the rectory. That person would have to be Mrs. Thomas West Williams, who moved to the T. W. Williams house with her husband around 1903 (A. Porcher).

My parents, Dr. J. Norman and Cleremonde Walsh, rented the rectory from 1936 until 1941 or 1942. I have fond memories of sitting on the front porch and listening to rain on the tin roof. There was no hot water, and baths were taken in a round tin tub. In winter, the tub was placed near a Franklin wood stove upon which the water was heated. Meals were cooked on a wood-fired stove in the kitchen, which, at that time, was separate from the house but connected by a walkway.

In 1939, the author was riding in this airplane built by his father, Dr. J. Norman Walsh, at the Episcopal Rectory.

While living in the rectory, my first nursemaid (about 1937) was Rosalee Perry, who, at age ninety, remains a friend. Following Rosalee was Susan Middleton, who taught me how to ride a bicycle even though she couldn't ride one. My father was a master woodworker and, in his spare time, built a swing set, swimming pool, and an airplane ride, which was a favorite among the children in the village. The airplane was suspended on a cable hung between two trees about fifty yards apart and was launched from a platform at the high end. It had a propeller that turned as it glided down the cable.

The last renters of the rectory, who subsequently purchased the home, were Edgar and Mary Dwight Davis and their children, Katherine "Kitty" Davis DeLoach, Marianne Davis Monroe, and Edgar "Eddie" Davis Jr. Eddie Davis now lives across the road on the property where F. A. Porcher built his second home in 1844. The rectory, as it is still called, was sold to Denny West, William Iselin, and then to the present owners, Amber and Don Weathers.

Pinopolis United Methodist Church

1833 Pinopolis Road (map: 8)

Methodists living in Pinopolis attended Trinity Episcopal Church for many years. Beginning as early as 1830, though, efforts were being made by some Pinopolis Methodists to have their own church. Unsuccessful attempts to establish a church were made in 1871 and 1886. For a while, members held services in the home of Margaret Evans Harvey (Mrs. Harlock Huxford Harvey, 1825–1904).

On August 13, 1897, Margaret E. Harvey's son, Joseph Arnold Harvey, purchased a one-acre lot from the estate of William Cain for $100 to be used for the site of a church. John Purkey Harvey, Joseph Arnold Harvey's first cousin, left $200 in his 1897 will to be invested and the interest used to pay a portion of the salary of a preacher. The *Southern Christian Advocate* on April 5, 1900, mentions that the church building in Pinopolis had been dedicated the previous Sunday. The first pastor was the Rev. W. T. Patrick (Pinopolis United Methodist Church 1986).

The Pinopolis United Methodist Church held its first service in March 1901.

Since a parsonage was needed, the Pinopolis Methodist Church purchased an additional one and one-half acres from Joseph and Calhoun Cain for $150 on September 10, 1901. In 1916, a church report showed that repairs were made to the parsonage for $36.65 and that its value had increased to $1,200. The old Methodist parsonage was torn down in 1956 and replaced with a brick dwelling. No photographs of the parsonage have been located; however, Anne S. Ball states that the parsonage was similar to the Grant house in appearance.

Ministers at Pinopolis Methodist were John Lawton Ray; W. H. Thrower; W. T. Patrick; W. E. Barre; C. W. Burgess; J. P. Winningham; W. H. Perry; W. C. Gleaton; J. W. Bailey; W. P. Bedenbaugh; J. E. Clark; Julius Franklin; J. W. Jones; Kayton; Spencer Carmichael; Robert Pinckney Hucks; George Harland Varn; B. C. Gleaton; Ben H. Covington; Willie Davis; Henry J. Bennett; Charles Polk; Henry Ford; R. M. Tucker; Charles A. Graves; Henry Thompson; John C. Snyder; Richard Quinlan; J. W. Jenkins Jr.; Robert Hopper; George Riser; Mickey Bell; William Childs; Frank McCoy; Robert Tanner; Mary Frances Lynch; James E. Alewine; Robert Lewis Bozard Jr.; and Christopher L. Poole Jr. (Pinopolis United Methodist Church 1986).

The present minister, Richard A. Broomall, began service on June 15, 2004.

St. John's Baptist Church

1142 Sugar Hill Drive (map: 1)

St. John's Baptist Church is located one mile south of the Tram Road, which was once considered the southern boundary of Pinopolis. While technically outside of Pinopolis Village, St. John's Baptist Church is closely linked with

Pinopolis residents. In 1884, J. W. Walling, A. D. Hare and A. Jones donated the property for the church. They stipulated that it was given specifically for a church and that the land would revert to Hare if the church were removed. Rev. D. M. Breaker and Rev. Samuel Lynes began the initial organization of St. John's Baptist Church and also adopted a church covenant and declaration of faith. Reverend Breaker was the first pastor of the church while William J. Dennis and John W. Walling were the first deacons.

Walling and Dennis submitted plans for a church building thirty by thirty feet with walls fifteen feet high, three doors, and fifteen windows. Two hundred dollars had been raised initially for contractor J. L. Driggers to complete the work, but final building costs ran to be $275. The historic marker on the road beside St. John's Baptist states that it was constituted in 1851 and the building completed in 1881.

The Moncks Corner Baptist Church opened in 1926, and many St. John's members transferred memberships there. In 1945, J. O. McCants deeded two and one-half acres of land adjoining St. John's Baptist Church to the trustees of the Moncks Corner Baptist Church to be held in trust for St. John's.

Although St. John's is no longer organized to serve a congregation on a regular basis, volunteer pastors and laypersons still conduct Sunday sermons and a study class. Occasionally, weddings and burials are held there, and many local family names can be found on the tombstones. The organ, donated by Jacob Carson, is still in service. The interior is unchanged, and kerosene lanterns still hang from the ceiling.

Inset: Interior of St. John's Baptist Church.

St. John's Baptist Church was completed in the early 1880s. It has not had an active congregation since the 1930s; however, volunteer pastors still lead Sunday services.

Old Nazareth Reformed Episcopal Church no longer existing

Wampee Drive (map: AA)

Most antebellum planters who owned summer houses in Pinopolis were Episcopalians. Before the Civil War, slaves belonging to those planters attended Episcopal services in plantation chapels and Episcopal churches. Services for slaves were sometimes held directly after the regular service; oftentimes slaves were allowed to attend the regular service with their master's family, although segregated into the balcony of the church. After the Civil War's end in 1865, Negroes were not welcomed into the Episcopal Church and were left homeless as far as their church needs. Although there are no records to verify the fact, it is likely that Negro Episcopalians in the Pinopolis area in 1865 found a gathering place of some kind to worship together.

In 1865 after the Civil War, Rev. P. F. Stevens, former rector of Trinity Episcopal Church Pinopolis, organized the freedmen into congregations. The term "freedman" referred to a slave who was now free as a result of the Emancipation Proclamation. Stevens is credited with helping the freedmen build their own churches and roadside chapels, which operated independently of the plantations (McCarrier).

Frank Ferguson of Pinopolis, a freedman in Immanuel Protestant Episcopal Church, was trained for the Episcopal priesthood and passed all required examinations. However, his application to be a priest in the Episcopal Diocese was rejected three times. Ferguson wrote in his own words about the early years of training and the formation of the Reformed Episcopal Church: *I served as Catechist in the Protestant Episcopal Church five years, from 1870–1875. On or about December 11, 1874, a Convention was held in the village of Pinopolis, consisting of delegates from Calvary Church in Charleston, Redeemer Church in Pineville, Nazareth Church in Pinopolis, and Bethlehem Church in Oakley to take into consideration what course they would pursue to bring about permanent organization among the Colored people who were Episcopalians* (McCarrier).

This building, labeled Old Negro Church, Wampee, 1895, could have been the first Nazareth Reformed Episcopal Church building. *Courtesy of the Cain files, South Carolina Historical Society.*

From Ferguson's above statement about a convention of Negro churches, it can be assumed that Nazareth Church had already been organized and was functioning as of the convention date of December 1874. Nazareth Church history states that Nazareth Reformed Episcopal Church had its origin in a tent on Somerset Plantation on June 6, 1875. The Negro Episcopalians of several churches, including Nazareth, gave up trying to be a part of the Episcopal Church and so joined the Reformed Episcopal Church in 1875. The first rector of Nazareth Reformed Episcopal Church was Rev. P. F. Stevens, who rejected his own religious background in the Episcopal Church and joined the Reformed Episcopal Church. This was quite a bold move on Stevens's part as he was breaking rank from his planter class to associate himself with the Negro congregation. Stevens later became a bishop in the Reformed Episcopal Church. Rev. Frank Ferguson of Pinopolis, who had trained for the Episcopal ministry from 1870 to 1875, became the first black rector of Nazareth Reformed Episcopal Church (McCarrier).

The Old Nazareth Reformed Episcopal Church congregation could have met as early as 1865. Construction date of the building shown here is unknown, but it was probably around 1895. The congregation moved to Moncks Corner in 1944, and the building burned in 1950. *Courtesy of the Berkeley County Historical Society.*

The first Nazareth Church building was erected at Wampee Plantation in 1877. The church shown on opposite page may have been the first Nazareth Church building. The Wampee Plantation land on which the church stood had not been deeded to the congregation, and so, fifteen years later, the congregation was forced to move its church. Richard Macbeth of Wampee Plantation gave Nazareth Church property for a new church on the southern portion of Wampee Plantation. This church, known as Old Nazareth, was built about 1895 and used until 1944 when the congregation built a new church in the West End area of Moncks Corner. Old Nazareth burned in 1950.

The first Nazareth Reformed Episcopal Church in Moncks Corner was replaced in 1968 and again in 1990. Those who have served in the pulpit have been Stevens, Ferguson, Bash, West, Capers, Grant, Madison, Smith, Mazyck, Liferidge, Collins, Blandon, Warren, Aiken, Harris, Rembert, Addison, Lloyd, and Milligan (Nazareth Church).

The site of Old Nazareth Church in Pinopolis was incorporated into the Old Nazareth and Wesley Cemetery, which remains in use today (map: 37). The cemetery is shared by the present Nazareth Reformed Episcopal Church and Wesley United Methodist Church.

This bell was removed from Old Nazareth Church Pinopolis and rests in front of the Nazareth Church in Moncks Corner.

Wesley United Methodist Church
(map: Ⓐ)

In 1800, a group of Christian slaves met in a house in the Bayfield section of Pinopolis. Under the leadership of the Reverend Harrison, they called their group the Wesley Episcopal Church. After the Civil War, the

The persons shown here are descendents of former members of Old Nazareth Church Pinopolis and are members of Nazareth Church Moncks Corner.

congregation joined the Methodist Episcopal Church. The Wesley Methodist Episcopal Church of Pinopolis was officially founded in 1866.

Twenty years after the Civil War, the congregation was able to build its first church building. In 1884, Jacob Carson and Harlock Webb Harvey donated land just south of the Tram Road for the church. Simon Gibbs donated lumber for the church building. The Wesley Methodist Episcopal congregation used this building until 1970 when a new structure was built closer to Moncks Corner.

Before the creation of Lake Moultrie, Wesley Methodist Episcopal Church had been a central location for many Negro communities surrounding Pinopolis: Hog Swamp, Bay Field, Kittfield, Green Hill, Whiteville, Wampee, Somerset, Sommerton, Broughton Hall, and Spann Grove.

Jingo Manigault (1830–1934) was fifty-five years old when the first Wesley Church was built. Jingo was a prominent figure at Wesley. He served as a sexton in the church but has become almost legendary as its bell ringer. Manigault walked over two miles from his home in Kittfield to Pinopolis every Sunday morning to ring the Wesley Church bell and wake the congregation. He then walked home and returned later to Wesley to ring the bell again, this time announcing the church service. Jingo Manigault's son, Charles Sinkler Manigault (1887–1967), became the next dedicated bell ringer. Wesley Church members came by wagon, buggy, and ox cart. Many also walked. Legend tells that some walked barefooted, carrying their shoes to put on just before entering the church.

This is the bell from the original Wesley United Methodist Church building in Pinopolis.

Jingo Manigault was the first bell ringer at Wesley United Methodist Church in Pinopolis. *Courtesy of Cynthia Hamilton.*

In the twentieth century, the church's name changed to Wesley Methodist Church after the merger of Methodist Episcopal Church South, the Methodist Protestant Church, and the Methodist Episcopal Church.

Rev. John D. Boone was pastor from 1958 to 1964. One of Boone's goals was to educate the members about the need for a new church building. Boone made the first contribution of $100 to back up his words. Rev. I. N. Stewart, who created a formal building fund program around 1965, followed Boone. Five years later, the proud congregation of Wesley Methodist Church walked from Pinopolis to its new debt-free building near the intersection of Highway 61 and Whitesville Road.

The present pastor of Wesley United Methodist Church is Rev. Robert L. Shuler (Wesley United Methodist Church).

Charles Sinkler Manigault took over his father's duties as the bell ringer for Wesley United Methodist Church in Pinopolis. *Courtesy of Cynthia Hamilton.*

chapter VI

Black Families of Pinopolis

Oh, Lawd, trouble een duh lan',
Yuh gottuh tek duh chillun
outuh Pharaoh han'

Negro Spiritual originating in Pinopolis
Smythe, et al. 1931

BEFORE THE CIVIL WAR, SOUTH CAROLINA LAW PROhibited teaching Negroes to read and write. It is not surprising, therefore, that very little written history has been found about black families living in St. John's or Pinopolis during that time. In 1865, the Emancipation Proclamation delivered the Negro from "Pharaoh's hand." Living conditions did improve for the former slaves and their families, but even in 1923, some fifty-eight years later, black families still struggled to survive. Mary "Dolly" Brown was a little six-year-old black girl living in Pinopolis in 1923. School conditions for Dolly and all black children were primitive. Her family was poor, and survival was the primary objective.

As I began interviewing people for this book, I realized that, although black and white families had lived side by side in Pinopolis for over a hundred years, they had not lived parallel lives. A white person in an interview for the book commented, "We didn't know the black people back then." A response from a black Pinopolis resident gave me a more telling explanation, "We were invisible." Many of the older black women I interviewed knew a great deal about me as a child and remembered many incidents about my life. I'm sorry to say that I could not recall the same details about their lives. That was just the way it was.

When I entered the county's public school system in 1941, I was a child and not aware of social issues. I was certainly not aware of educational inequalities between whites and blacks.

Dolly Brown's daughter, Maude Smalls, is my age, and we both grew up in Pinopolis. I asked Maude how she got to school when she was a child. She explained to me that, as the school bus picked me up for my ride from Pinopolis to Moncks Corner, she began her four-mile walk to Moncks Corner for school. School busses were not available for black children until 1945. Maude told me that Ed Beatty, a neighbor of theirs, had frequently taken her to the R. A. Ready School at the east end of Main Street on Highway 52. On days he didn't, Maude walked into Moncks Corner and back every day.

The black people I interviewed for this chapter of the book were raised in Pinopolis or had parents who made their homes in Pinopolis. Many were the descendants of slaves who were owned by the founding families of St. John's Berkeley, more specifically, families of Pinopolis. After the Civil War, their livelihoods changed in many respects, but they were still dependent upon their former owners for employment. Their churches played an integral role in their personal lives, as did their relationships with their employers. Black families contributed greatly to the community feeling in Pinopolis. I was happy to listen to their stories and to "see" them for the first time.

Mary Rebecca Sheppard Brown, known as Dolly, was born at Bearfield Plantation in 1917. Dolly's grandmother and grandfather, Mary and Tony Porcher, raised her. Tony Porcher worked as a handyman. Henry R. Dwight Sr. owned and farmed Bearfield at that time. Henry Dwight Jr. recalled that his father built four tenant houses at Bearfield Plantation for black families who worked there. Each house had only two rooms, and Dwight wondered how Dolly and the other families were able to live in such small spaces.

Nina Dwight (1909–1991), daughter of Henry R. Dwight Sr., and her sister, Mary Macbeth Dwight (1905–1983), taught Dolly to read. They made a little school in Dr. Dwight's office building in Pinopolis. Before this schooling, Dolly spoke only Gullah. Nina Dwight used two blackboards, writing Gullah words on one and English on the other. By comparing the two, Dolly learned English.

Mary "Dolly" Brown was born in Pinopolis and has written about some of her experiences there.

When Bearfield Plantation was to be flooded by Santee Cooper's creation of Lake Moultrie, Dolly and her family moved to a little house near Old Nazareth Church in Pinopolis, which they rented from Elizabeth Cain (map: △2). In 1938, after Dolly moved to the village, her first husband, McKinley Smalls, was killed by an automobile on Broughton Hall Road. She now lives next to the new Wesley Methodist Church. Her daughter, Maude, lives nearby and cares for her daily.

As far as I know, Dolly Brown is the only one of the black people who lived in Pinopolis who has written a memoir. Despite having only a third-grade education, Dolly's memoir captured her memories well. Dolly called her story "History About People and Places." The version printed here has been slightly edited.

On August 8, 1917, a small girl was born to a good and wonderful couple. One of six children, she was named Mary Rebecca Sheppard. At the age of 6 she went to the family of Mr. & Mrs. Dwight in Pinopolis to just bring wood to fill the box behind a big wood stove, and fill the pail that sat on a shelf with water from a pump. Oh well, I would take my hand and let some run into it and drink. I hope none didn't get in the bucket. That I wouldn't remember because it has been a long long time—many moons ago. Then Mrs. Dwight had long hair and she would let me scratch & brush it. Sometimes when I scratched I thought that I was doing my doll's hair. Very softly she would say to me "Dolly go easy now." "Yes ma'am" would be the answer. Then we would talk about the family—how hard everybody had to work in the fields. Mr. Dwight owned a place called Bearfield and I lived in one of his houses. That is how I got to know the Dwight family. All the tenants that lived in the houses worked on the farm and plowed the fields. They worked around the Dwight home cleaning the yard, cutting grass, hoeing the garden, sweeping the yard, cooking in the kitchen and lots more. We never had any welfare, food stamps, social security—nothing but the love of God deep down in our heart—that was the thing that held us together. I was one of the lucky ones that got a chance to be able to get to work at that family. Their youngest daughter was Mrs. Nina Dwight Langley. She would take time out to teach me how to read and write. I went to school and one day Nina & I went up in the old school house [the Pinopolis School] *and rang the bell. People came looking for a fire. That was the last bell we ever rang.*

This typical Negro tenant home in 1939 shows newspapers that were often used for insulation. *Courtesy of Jack Delano, Santee Cooper.*

Please don't forget the three stores: Mr. Wiley, Mr. Cuthbert Hare, [and] *Mr. Harley Harvey. There were three good stores in Pinopolis. The candy was real good: Baby Ruth, Jaw Breaker, Banana Wonder, Licorice Stick, Mary Jane, All Day Sucker, Peanut Bar, and so much more. When I think about the good times we had in Pinopolis and how everything has changed, I could cry.*

One day Mrs. Betty [Elizabeth Ravenel Cain Dwight, 1877–1929] *said to me "Dolly, take the broom and brush the cobweb from the chandelier over the dining room table." I wanted to go and play and was not paying any attention to her. In those days the table had white linen cloth on it, beautiful white as snow. I got the broom and pulled up a chair. I got on the table with dirty bare feet and walked all over the table making black tracks. I looked and the door opened. There stood Mrs. D. amazed at me on the table and her nice cloth covered with dirt. But she did not frighten me but spoke softly: "What are you doing?" "Ma'am I am brushing down the cobweb." That was one job I never had the chance to do again.*

For some 30 years I worked for the Massalon family in Pinopolis. Everyone treated each other with respect. I wonder what has happen to the families now. Not so close anymore—it is sad, sad, sad. We as kids in Pinopolis all would swim in the big swimming hole. Never thought about snakes or nothing. So I see now that God takes care of us because we didn't know any better. Thanks be to God for his great love and mercy. Oh how I remember the good old times.

Wiley's store was one all the kids used. He gave credit to buy candy and popcorn and he would let us have it. But pay up and don't come for more before you pay for the first ones. We didn't always pay, so one night he scared our hearts out of our body—chased us, sent a few shots over our heads and that was end of the credit for us! But we still loved him for straightening out things for us. Pay first; get more later.

I hauled milk from Mr. & Mrs. Carson and it was fun because I looked forward to playing with their two girls, Carolyn and Mickie Carson.

I also had the pleasure of working for Mr. Herbert Thomas, owner of the Pepsi-Cola Company from North Carolina. The most exciting of all was the General Mark W. Clark at the lodge on the lake in Pinopolis. I fixed biscuits, fried grits, fried fish, and cheese toast. That was one thing he liked for me to fix. Dessert was apple cake with Cool Whip (Brown).

Catherine Beatty (1894–1994), called Dove or Kat, and her husband, Ed Beatty, lived near Broughton Hall Road (map: △4). She was a longtime cook for Anne and Doc Fishburne. The Fishburnes' granddaughter, Anne Moultrie Ball Helms, known as Moonie, had lunch with them every weekday and thought of Kat as a second mother. Ed and Kat had no children, but they raised Laverne Beatty Mazyck, Ophelia Beatty Morris, Sandra J. Beatty Morris, William James Beatty Morris, and Adrian Shine. Living on the Beatty home site now is Ophelia Morris.

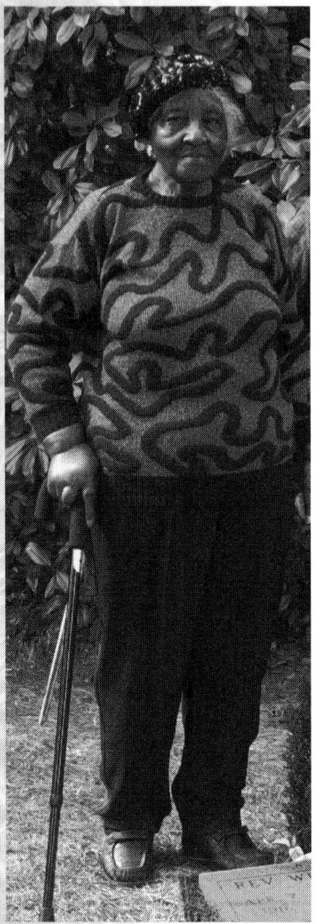

Rosalee Ferguson Perry lived and worked in Pinopolis for many years, beginning in 1937.

Several families lived on the Fishburne property (map: 24). Louisa Johnson, known as Babe, took care of Doc's dogs. When Louisa's house burned, Anne Fishburne immediately remodeled another building for her home. Frank Simmons, who was single, was the gardener. Lucinda Thompson, called Dee, was married to Franklin "Buster" Brown, who managed Dr. Fishburne's plantation, Foxbank. Dee and Buster lived on the grounds of the Fishburne's Pinopolis home (map: △7). Dee worked for Anne Fishburne Ball and helped to raise her daughter, Moonie. Dee's daughter, Lou Ann Brown Smith, was a contemporary and playmate of Moonie. After graduating from Claflin College, she worked in New York City until her retirement in 1994 when she came back to live in the home her parents built near St. John's Baptist Church.

Rosalee Ferguson was born November 19, 1915, at Eadytown. She went to school there through the sixth grade. She came to Pinopolis in 1937, lived in a small building on the rectory lot, and worked for my parents, Dr. Norman and Cleremonde Walsh, from 1937 until 1938 when she left to take a job in Knoxville, Tennessee. Her cousins, Carrie and Florence Richburg, who had also worked for the Walshes, recommended Rosalee to them. Rosalee met William Perry, who was working for the CCC (Civilian Conservation Corps), and married him in 1939. They moved to Bearfield and lived there until flooded out by Lake Moultrie. Bearfield was near the Chapel Hill summer house. They lived for a short time in the Nazareth School House next to the church (map: Ⓩ). About 1942, William built a house in the junction of Broughton Road and Wampee Drive (map: △3).

Rosalee next worked for the Dwight Porchers. Most of her work life was "minding children and doing housework." William worked for a while as an assistant surveyor and later at the Charleston port. After that, he became a minister in the Reformed Episcopal Church. Their first parish was Cainhoy. Rosalee has two daughters, Cynthia Bryant and Dianna Savage. Their Pinopolis home burned around 1985, and they moved to Moncks Corner. William died, and Rosalee's grandson, Robert Perry, and his family now live with her. Names she remembers from Bearfield are Willie Gethers, Louise Tucker, Eugene Heyward, and Lizzie Poinsett. J. P. Wilson lived near the Old Nazareth Reformed Episcopal Church cemetery.

Emily Palmer (1933–2006) grew up in Pinopolis but now lives in Whitesville. Her parents were Robert and Eliza Palmer. The first home she remembers was near the Old Nazareth Church in Pinopolis. She went through the third grade at Nazareth School and was taught by Verbatine Hamilton. After that, she went to the Berkeley Training School in Moncks Corner. She walked to school until the fifth grade when a bus was provided. Emily's siblings were Francis Palmer, who died in 2000; Catherine Palmer, who died in 1951; Emmatrude, who died in 1972; Thaddeus Perry Palmer; and William Perry. Emily died several months after this interview.

This 1939 image shows the porch of the same Negro tenant home shown on p. 139. *Courtesy of Jack Delano, Santee Cooper.*

Missouri Middleton, her daughter, Stay, and her sister, Susan Middleton, lived on the Chapel Hill Plantation summer house lot. Missouri worked for the Percival Porcher family, and Susan worked for the Walsh family.

Essie Major Burch and her husband, Frank, lived for a while in a house near Jacob's house (map: 8). Their daughter, Isabella Brown Thomas, now lives in the Green Hill community near the West Dike. This area is south of the Tram Road and west of Pinopolis Road. Isabella Thomas's sister Ellen worked for Connie Harvey Murray and cared for Connie's daughter, "Dinks," who married Carlton Morrow. Then Ellen took care of their children.

Black families who were relocated when Lake Moultrie was created settled Green Hill community. According to Albertha Brown Varner, daughter of George and Elsie Washington, Brutus Green owned a large tract of land where the community is now. Much of it is in the category of "heirs' property" where clear title is difficult to obtain. Present-day residents of Green Hill agree that their community is a part of Pinopolis (interview with author May 12, 2006).

Eight negro families were settled on eight acres of land near Moncks Corner in 1939. *Courtesy of Jack Delano, Santee Cooper.*

Planting crops by hand in 1939. *Courtesy of Jack Delano, Santee Cooper.*

Negro farmer in 1939. *Courtesy of Jack Delano, Santee Cooper.*

Negro farmer's wife in 1939. This was just before Santee Cooper began relocating families, and she may be taking a last look at her home. *Courtesy of Jack Delano, Santee Cooper.*

chapter VII

Education in St. John's Berkeley and Pinopolis

"I thank God there are no free schools nor printing, and I hope we shall not have these [a] *hundred years* [from now]*; for learning has brought disobedience and heresy and sects into the world, and printing has divulged them, and libels against the best governments. God keep us from both."*

Sir William Berkeley, 1671 (Cross 1985)

ACROSS THE ATLANTIC, THE LORDS PROPRIETORS of South Carolina, highly educated Englishmen, were well aware of the advantages—and disadvantages—of education for the children of the colonists. Sir William Berkeley's quotation expresses how important it was for English rulers to keep the common person uneducated. Early on, people of means, including St. John's Berkeley planters, were concerned about the education of their children. They were well aware of the value of a proper education. The ideal man needed training in Greek, Latin, and the classics. He took a responsible role in government and the church, and he had enough scientific interest to learn the best farming techniques. His education should be shaped around these needs.

Young children usually received educational instruction from female members of the community who taught children from several families as well as their own. Live-in tutors, some from as far away as New England, were also employed. Daughters were seldom schooled beyond the basics of reading, writing, and arithmetic. Young ladies were taught proper running of a household and social graces. Well-to-do families sent their daughters to Charleston for finishing touches in etiquette.

Many in the upper classes recognized the need to educate citizens from all classes. Private donations were being made toward public education in the early 1700s.

It would be another two hundred years before the state government assumed responsibility for public education in South Carolina. In 1728, Landgrave Thomas Smith of Yeamans Hall gave forty acres of land for a free school in Wassamassaw Lower St. John's, naming as trustees Isaac Porcher, Cornelius Dupre, Francis Cordes, and Abraham DuPont (Cross 1985). Funds were raised in St. James Goose Creek Parish to educate poor children, and in 1828, the vestry of St. James Goose Creek Parish established a school there.

Lack of education in the "back country" was a chronic issue: *There was such a lack of schools prior to the American Revolution that in January of 1768, the grand jury presented as a grievance "the want of schools in the back parts of this province, for the education and instruction of the poor people who are settled there, and the want of a law to hinder unqualified people preaching throughout the province"* (Orvin 1973).

As early as 1767, records show a school in the Pineville area. Later records show George King, a Pineville planter, keeping a school on his farm (Cross 1985). Pineville was settled in 1794. By 1805, Pineville Academy had been incorporated. Well-known for academic excellence throughout the state, Pineville accepted both boys and girls. It was not unusual for Pineville Academy graduates to continue their education at places like Yale and Princeton. Pineville Academy prospered under the leadership of its first schoolmaster, Mr. Alpheus Baker from New Hampshire, who served from 1805 until 1812. After Baker left, the Board of Trustees advertised for a principal at a salary of $1,000, which was to include a residence. They also advertised for a teacher at a salary of $800, residence included.

In 1818, Frederick A. Porcher, age nine, attended Pineville Academy. Porcher boarded during the school week with one of his teachers, Mr. Charles Stevens. In 1820, Henry William Ravenel, who was to become a well-known botanist in South Carolina, began his education at the Pineville Academy. In 1836, Pineville Academy closed for several years because of the malaria epidemic (Orvin 1973). The academy probably reopened around 1840 and closed permanently after the Civil War when the village was burned.

In 1811, the South Carolina General Assembly authorized the establishment of public schools in every district in the state but provided little funding. In 1860, only about half of the state's white children were in school. During Reconstruction, the state's black leaders worked to establish a publicly funded system of education for all children, white and black (Edgar 1998).

The Freedmen's Bureau also set up schools in South Carolina for recently freed slaves. One such school operated in Pinopolis for a few months in 1868 before moving to Moncks Corner. The Freedmen's School in Pinopolis had one white female teacher, Miss Mary Von Hagen. There were forty-six male and nineteen female students (District Superintendents Monthly School Report, South Carolina, January 1868).

Another school for Negroes, the Pinopolis Nazareth School (map:Ⓩ), began operation sometime after 1883. Gen. E. J. Dennis surveyed the lot for the school and drew up the plat in 1883. The Nazareth School was operated by Berkeley County and served the Pinopolis community until the late 1930s. After the school building was no longer being used, various black families lived in it. It was later demolished.

In his book *Slaves in the Family*, Edward Ball states that Henry Martin, who was born a slave on Limerick Plantation, attended Nazareth School in Pinopolis in 1866 and later taught there. Local histories of the Nazareth Church and School do not mention Peter Henry Robards Martin. Ball's statement is difficult to verify because it is uncertain whether Henry Martin ever attended school in Pinopolis. The Freedmen's School operated for only a few months in 1868, and the Nazareth School did not begin until the 1880s. It is certain, though, that Henry Martin did live in Pinopolis. He wrote letters from Pinopolis to friends in Charleston, including his former master, Isaac Ball. Martin became a minister, teacher, and provider for orphaned black children. Henry Martin maintained a lifelong friendship with the Ball family. Martin expressed his philosophy about slavery in this way: *My father's people were never free, and they were never slaves, so far as the word slave may mean* (E. Ball 1998).

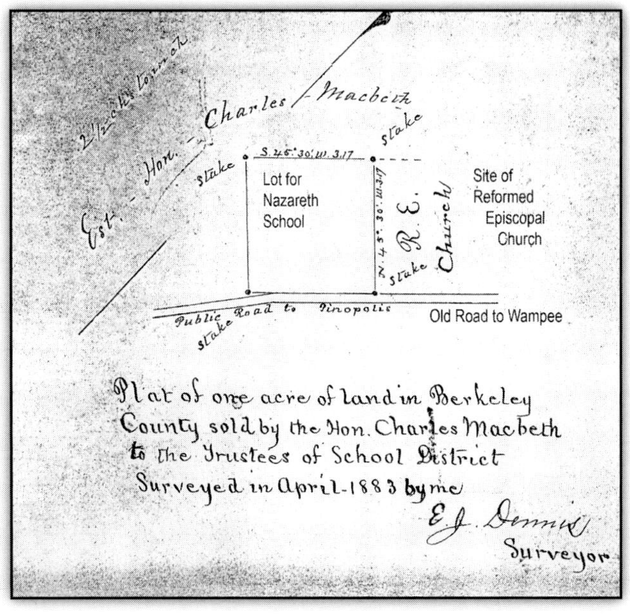

This plat shows the location of the Nazareth School in Pinopolis.

M. C. Orvin notes that Berkeley County citizens in 1886 were concerned about the inadequacy of public education. A Berkeley County resident deplored the situation in a letter to the Charleston *News and Courier* on April 5, 1886: *The citizens of our county anxiously watch the progress of the pending educational bill. The illiteracy of Berkeley surpasses that of any county in the state. It is verily the "Black County," physically, mentally, and morally. Its intellectual status warrants its name, "Baby County," and the meager yearly appropriation of $40 per school supplies but a scanty educational pabulum. Settlements are scattered over the face of the county where hundreds of white children—the future rulers of this great Republic—were never taught a letter of their mother tongue. The county is inadequate to its task, and without a speedy help from an outside source, the prospects are anything but hopeful.*

The meager remuneration of our teachers leads men of ability and competence to seek a new field of labor and leaves us but a scanty number of half-educated teachers who perform their task simply as a matter of duty. Like pedagogical tramps, they traverse the county in search of employment, learning little, teaching less. The youthful mind, like a frail bark, is tossed on the sea of confliction theories and text books until it hopelessly settles into a state of comparative imbecility. The greatest drawback on the educational prospects of this county is the so-called "license system" by which parties void of all knowledge assume the responsibilities of tutors. This system renders the examinations before the county boards of examiners a farce, and opens a broad field for unworthy speculations. These "licensed teachers" are scattered all over the county, endeavoring to impart to their pupils the knowledge they themselves fail to possess.

As a partial answer to this concern, Berkeley County opened a training institute for Negro teachers in Mount Pleasant in July of 1886. Thirty teachers were registered (Orvin 1973).

Information about education in Pinopolis from its settlement in 1834 until the formation of the privately funded Pinopolis Academy in 1895 is scant. Private diaries, journals, and letters show that schooling took place in the homes of Pinopolis families like the Cains, Macbeths, and Ravenels.

Marie Jervey Ravenel, in her book *From Childhood to China*, described her childhood in Pinopolis in 1898: *I was frantic to go to school. Finally, when I was about six years old, I was allowed to go to my grandmother's school. The first time I got ready to go with the crowd over to school, I just thought I'd pass out with pleasure. And it was right next door* (M. Ravenel 1991). Maria's home was the Rene Ravenel Jr. house (map: Ⓣ). To the west was the Pooshee summer house (map: 27), home of her grandparents, Isabelle Deveaux Ravenel and Dr. Rene Ravenel Sr.

Maria's grandmother's home, like many others, had an outbuilding called "the lodge," which could be used to house extra male guests or family members, but in this case, it was used as a school. *My only fear was that they always had a cow in the lot between my house and the school and I was afraid of that cow. Grandmother had started the school because there was not another one there at the time and she wanted to take care of her grandchildren and the other children in the village who didn't have governesses. The lodge behind my grandmother's house had two rooms, one where Uncle Percy slept and the other one which she used as a classroom. As far as I can remember, there were about ten to twelve children in the school. I doubt that my grandmother had had any teaching instruction, but she taught us reading and writing and arithmetic. Those were the main things, but we didn't get much more. In those days before there were any public schools in Pinopolis, after the boys finished studying on the elementary level, they all had to go off to continue their educations . . . to Porter and then to Clemson and Charleston College.*

After that, [a] *public school was established, run by a board who hired a teacher. I remember when they built it . . . we called it "the big school"* [Pinopolis Academy]. *We went there and it was just for Pinopolis children . . . there wasn't anybody else. The school was built near a little stream which we used to wade in and ruin our shoes and get punished . . . we didn't get our Uncle Remus when that happened.*

Sketch of the Pinopolis Academy by Jan Dennis.

The Pinopolis Academy

(map: 18)

After a while [after the Pinopolis Public School had been built], *the Moncks Corner people began coming in because there was no school over there. They either came the few miles in a horse and buggy but mostly they walked on their two legs. But they could only come for half the year. As soon as it was time to plant, they would have to stop and stay home . . . they all worked in the fields. I know our cousins the Cain boys had to walk the two miles from Somerset Plantation to our school* (M. Ravenel 1991).

Minutes from the meetings of the Pinopolis Academy Board of Trustees contain a great deal of information about Pinopolis Academy's history. The minutes tell us that on December 10, 1895, five men incorporated the Pinopolis Academy and became its first trustees by contributing capital funds: Dr. Joseph P. Cain ($100), Thomas W. Williams ($100), Rene Ravenel ($200), J. B. Morrison ($100), and John Purkey Harvey ($100). Their purpose was to fund, erect, and conduct a school in Pinopolis for boys and girls from primary to advanced grades.

On May 20, 1896, Joseph P. Cain and J. Calhoun Cain, as representatives of the estate of William Cain, gave the Pinopolis Academy title to the property in a document sworn before Magistrate Robert R. Lindsay. Virgil C. Dibble was hired as the first principal on July 1, 1896, at a salary of $550 per year.

There are inconsistencies between the minutes from the Board of Trustees meetings and other histories as to when the Pinopolis Academy building was completed. Orvin says the school opened in January 1896 while Anne S. Ball says September 1896. However, it is likely that instruction took place at several different locations before building actually began. Minutes from April 17, 1897, show that on that date the Board of

The Pinopolis Academy building forms the central portion of this home built by Kinloch Bull in 1951. The Academy building was likely completed in 1897. Classes began in September 1896, possibly in private homes. The first commencement was held June 27, 1898.

Trustees gave approval to the construction of a new school building.

In October 1897, the Pinopolis Academy Board agreed to assess Arthur Hamilton DeHay (1857–1931) $40 for the year to educate his five children. At the February 26, 1898, meeting, the board dealt with a complaint from Mr. DeHay about being compensated for the lost instructional time since his children did not attend school during Christmas vacation. DeHay was given credit of two months tuition but was still dissatisfied and threatened to remove his five children from school.

Apparently, Mr. DeHay let his children stay on at the Pinopolis Academy because, on June 8, the board dealt with the problems of Mr. DeHay's son, Robert Edward DeHay (1883–1940), who was suspended for disrespect and violation of the rules of the Academy. Robert refused to properly reply to the principal and, at the direction of the principal, refused to remove his hat. After repeated refusals, he received a whipping. Mr. DeHay protested, but Robert was suspended for five days and was required to apologize to Mr. Dibble.

On August 5, 1899, the Pinopolis Academy Board of Trustees met with G. L. Riggs, who was Berkeley County's superintendent of education, and the county trustees for School District 10½, Richard Yeadon Dwight, MD, and John Henry Porcher. An agreement was reached to make the Pinopolis Academy a public school so long as the funds were available to pay the teachers.

At that August meeting, the Pinopolis Academy Board approved loaning out the schoolhouse for dancing purposes but required that a matron be present at all such functions. Minutes from the meeting also read that a course of study was approved: *Every class shall include Mathematics, English, Physics, Latin and one other language, French, German or Greek. For advancement to the next higher class, a percent of 75 is required.*

Rules of conduct were established for behavior, including a demerit system: *The Principal shall be empowered to administer corporal punishment—grounds—fighting in or around School House and grounds.* [Demerits or punishment shall be given for] *blasphemous and lewd language, insolence, and failure in recitation.*

On May 20, 1901, Isaac Porcher of the Berkeley County Board of Education advised the Academy Board that county funds had been exhausted.

The board agreed to an assessment of six dollars per pupil so that the Academy could continue to operate.

By 1910, a larger building was needed, and county funding was available. Because of a lack of space, some classes may have been taught in private homes. Julia Skinner related that she started the first grade in 1910 and that her class met in Margaret Deas Ravenel Porcher's home, the Woodboo summer house. It was probably during this time that Stephen Deveaux taught in the back room of the old post office.

The new school building (map: Ⓛ), designed by Henry R. Dwight Sr., was added onto the front of the academy building (map: 18). When the school began operation around 1911, it became known as the Pinopolis Public School. Children from Moncks Corner and other communities attended.

The Pinopolis Public School was built in 1911. *Courtesy of Martha Porcher Ravenel.*

Pinopolis Public School
(map:Ⓛ)

The school and the community used the auditorium, the former Pinopolis Academy building, for piano recitals, plays, and dances. Occasionally, Anne Sinkler Fishburne and Virginia Massey provided entertainment.

The Pinopolis Public School had two rooms on each floor. Beatrice Dennis Plyler, age ninety-four, remembers that the first, second, and third grades were located downstairs in the room on the right. The fourth-, fifth-, and sixth-grade classes met in the room on the left. Upperclassmen met their classes upstairs. Outhouses for boys and girls stood at the rear of the building on either side. Water came from an outdoor hand pump.

Bea Plyler remembers sharing an inkwell with her desk mate, Emily Fishburne (Whaley). When Mary Causey, her first grade teacher, married, Eleanor Brown Edwards took her place. Bea would occasionally go to Mrs. Edwards's home, the Chelsea summer house, for coaching with phonics and the multiplication tables. (Interview with author 2005).

The new school building had a bell tower atop the building. The bell rang to signal the opening and closing of the school day, as well as recess. When rung outside of school hours, the bell became the village fire alarm.

In 1924, the Pinopolis Public School downsized and offered only first, second, and third grades. The school closed completely in 1928, and students were bussed to Moncks Corner. The school building was dismantled in 1938, and its materials were used to build the Berkeley County Library in Moncks Corner. When I was a teenager, the old Pinopolis Academy building was still being used for dances. By 1950, it was abandoned and was surrounded by bushes and vines.

Kinloch Bull purchased the old academy and property about 1951. Later, he described how he acquired the property and built a home for himself and his wife, Julia. *In 1948 I moved to Pinopolis and started looking for a lot on which to build a home. I wondered if the former Pinopolis Academy could be bought. With the help of a local lawyer, and invaluable help from Miss Caroline Cain, I was able to secure signature on thirty-five deeds from Cain heirs from various parts of the country from Georgia to Canada. The amount paid the Cains by me for the property was $750. The lot had been neglected for many years and was completely overgrown with bushes, small trees, and vines, some up against and into the house. In 1951, I started working on the house on Saturdays and after work. The house itself was 40′ x 20′, and the sills under the house forty feet in length were hand hewn, as were the joist and studs. And all were of "red heart" pine. In the more than five years that I worked on the house I was able to put in an upstairs, add seven feet to the length on the east, and also porches to the north and south. I was able to get a number of bricks from a lock of the old Santee Canal. These are on the ground in front of the porches* (K. Bull).

Rev. Hal West and his wife, Elliott Bishop West, purchased the property from Kinloch Bull. They sold the home and lot in 2005 to the current owners, Steve Buckler and Susan Hillis.

Pinopolis Academy

OF SOUTH CAROLINA.

THIS FIRST GRADE CERTIFICATE

Is awarded to Henry Lucas for Proficiency in Recitations. Propriety in Deportment and Punctuality in Attendance during the Week past.

PERCENTAGE REQUIRED, 95. acquired 94 6/7

Principal. C. S. Hay Asst.

Wk ending Nov 22, 1901

WALKER, EVANS & COGSWELL CO 210028

A first grade certificate from the Pinopolis Academy. *Courtesy of Camila Bishop Henry.*

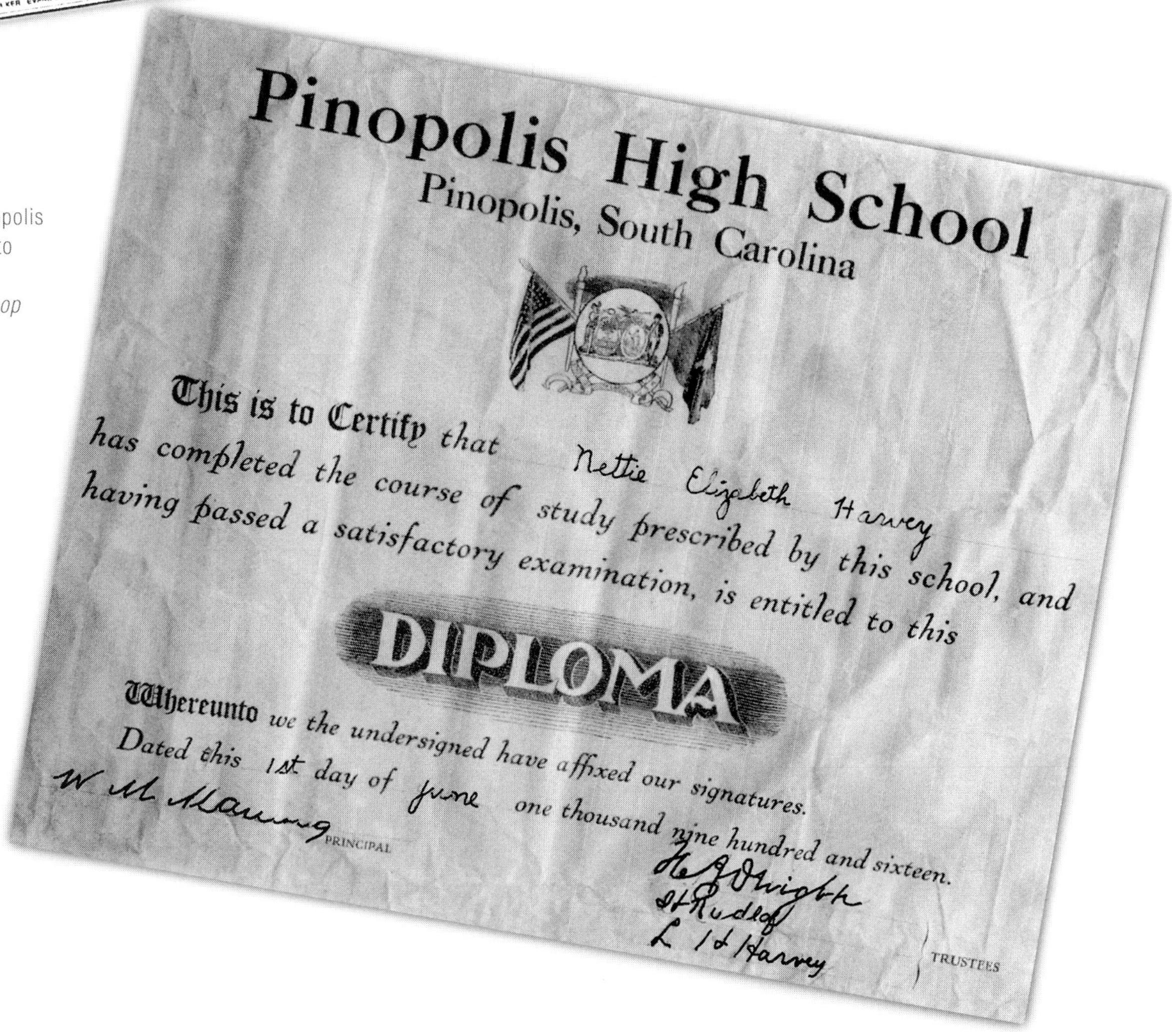

Pinopolis High School

Pinopolis, South Carolina

This is to Certify that Nettie Elizabeth Harvey has completed the course of study prescribed by this school, and having passed a satisfactory examination, is entitled to this

DIPLOMA

Whereunto we the undersigned have affixed our signatures.

Dated this 1st day of June one thousand nine hundred and sixteen.

W M Manning PRINCIPAL

H A Dwight
H Rudlof
L H Harvey TRUSTEES

A diploma from the Pinopolis Public School awarded to Nettie Elizabeth Harvey. *Courtesy of Camila Bishop Henry, her daughter.*

chapter VIII

New England City (Chicora), the Old Tram Road, and Pinopolis Cotton Gin

Should New England City become the county seat, in a very short time you will see erected not only a hotel, but a cotton factory as well as other factories that Berkeley may be proud of.

Albert S. Emerson

PINOPOLIS VILLAGE PLAYED A KEY ROLE IN THE RACE for political power at the turn of the century. Col. Charles Henry Simonton (1829–1904) introduced a bill to the state legislature in January 1882 to form a new county, Berkeley, out of Charleston County. Simonton's purpose was political—to separate the county along party and racial lines, keeping old Charleston Democratic. Most of the residents in St. John's Parish were Negroes and would continue to vote Republican. It was not until eleven years after Colonel Simonton's proposal, in December 1893, that a bill to create a new county passed and boundaries were drawn. A contentious debate about a site for the Berkeley County seat dragged on from 1889 to 1895. A group of Democratic politicians met in Moncks Corner in January 1889 to discuss the new county seat and suggested that Pinopolis, being in the center of the county, be selected. J. Calhoun Cain, a resident of Pinopolis, vigorously opposed this proposal because he and other Pinopolis residents did not want any development or commercialization of the village. In March 1894, once again the site for a county seat was brought up and the suggestions were New England City, Pinopolis, Macbeth, and the Barrows. In December 1894, Capt. Joseph Arnold Harvey and Capt. C. W. Sanders proposed Moncks Corner as county seat. Albert S. Emerson campaigned vigorously for New

A postcard from Chicora from the early 1900s. *Courtesy of Robert Austin.*

England City to be selected as the new county seat. He boasted in the *News and Courier* that New England City had an artesian well that yielded a bountiful supply of water: *the best mineral water in the United States . . . and that during the preceding four years not one case of fever occurred among those residents who took care of themselves and drank the water . . . if the people of Berkeley will locate the Court House at New England City we will give them a square for the Court House, a square for the jail, fifty acres of good, cleared arable lands and fifty acres of wooded lands for the poor house farm, and one-fourth of the proposed building material to be used in the construction of necessary buildings thereon, and, if the offer is accepted, will guarantee to furnish the balance of the material at reduced prices, and guarantee the completion of the Berkeley Railroad to Eutawville, and perhaps to Summerville . . . should New England City become the county seat, in a very short time you will see erected not only a hotel, but a cotton factory as well as other factories that Berkeley may be proud of* (M. Orvin 1950).

Emerson's campaign was unsuccessful. At an election in July 1895, the voters chose Moncks Corner as the new county seat. The next month, county records were transferred from Mount Pleasant to Moncks Corner (M. Orvin 1950).

Although Moncks Corner became the new seat of Berkeley County, Albert S. Emerson's New England City had been a leading contender. Emerson, a Boston entrepreneur, purchased property about eight miles northwest of Pinopolis, formerly known as Ward's Plantation and part of Ophir and Sarazins plantations. He intended to attract investors and residents from the North with stories of fertile farmland and a healthy climate. In August 1892, Emerson had a plat drawn with the title, "A Plan of New England City, Situate in the County of Berkeley, State of South Carolina, Thirty Miles from the City of Charleston" (Cross 1985). Essential to the development and growth of New England City was the creation of a rail line from Moncks Corner to New England City. In December 1892, New England City and the Berkeley Railroad Company were incorporated by the South Carolina legislature.

The Northeastern Railroad, later known as the Atlantic Coastline Railroad, had been running a passenger line from Charleston to Moncks Corner since May 5, 1855 (M. Orvin 1950). A. F. Emerson, president of New England City Corp., conveyed $75,000 to the Northeastern Railroad to connect a rail line from their tracks at Moncks Corner to New England City, a distance of ten miles. Emerson had hoped for the tracks to pass through the center of Pinopolis; however, the citizens of Pinopolis did not want the railroad to pass through the village, just as they had opposed Pinopolis becoming the new county seat. The tracks were laid in 1892, just north of the Harlock Webb Harvey store. The site where the railroad crossed Pinopolis Road became known as Pinopolis Crossing where a small station house (map:Ⓓ) was built. The track of the New England City rail line was accepted as the unofficial southern boundary of Pinopolis. Later the rail bed became known as the Tram Road after the line was sold and the tracks dismantled in 1907.

The train made two trips daily between New England City and Moncks Corner. Around 1900, the railroad was extended from New England City to Eutawville (M. Orvin 1950). The only passenger coach was an open platform car that had been bought from the Savannah and Isle of Hope Railroad in Savannah, Georgia. David Tyler (1894–1971), the son of New England City prominent businessman John Tyler (1857–1921), remembered that the fare from New England City to Charleston was twenty-five cents (Tyler). John Tyler moved from Cross to New England City in 1895. He was active in the thriving town where he developed many enterprises. He was the first postmaster and served for fifteen years. Tyler had a large mercantile business serving the local farmers, which was rebuilt and enlarged after a fire. He owned the Standing Yellow Pine Company. He farmed at Chicora and Ophir and had a cotton gin at Chicora (Cross 1985; Tyler).

The Chicora postmark shows that, at that time, letters were stamped at the destination. *Courtesy of Robert Austin.*

By 1898, New England City was being referred to on plats and deeds as Chicora. David Tyler moved to Chicora in 1895 with his parents. Tyler described Chicora in an unpublished paper found in the Elias Bull Collection of the South Carolina Historical Society: *I was born 9/12/1894 near Friendship Church* [Cross] . . . *we moved to Chicora when I was one year old. I can remember from when I was older that . . . there were pecan trees, offices, a bottling plant, a large sawmill, a schoolhouse, a Church, and a dance hall . . .* [There was] *a well 496 ft. deep . . . the water supplied all of the homes by pipeline, as it had a great pressure.* In 1930, long after Chicora had been abandoned, the well was still flowing.

David Tyler and his sister, Zadie (1888–1934), went by horse and carriage eight miles from Chicora to Pinopolis for school—"rain or shine." Sometimes, if there were snow and sleet, he and Zadie rode the train to Pinopolis. He remembered a depot at Pinopolis Crossing, near the Dwight and Cain cotton gin.

This image shows the Tram Road as it is today.

Tyler's memoirs are filled with names of people of the area between Cross and Pinopolis, including Sidney and Alma Crick, who lived near the railroad at a spot called "Crickville": *My contemporaries* [Tyler's] *would remember Mr. and Mrs. Sidney Crick. They lived initially at Chicora and then moved to Pinopolis on the property on the west side of Pinopolis Road just north of Hyrne's curve.* Mickey Owens, a longtime resident of Pinopolis and a neighbor of the Cricks, wrote: *The Cricks were an unusual couple, to know them was a treat . . . they met in Canada where both worked for a medical institution . . . they first came to Berkeley County in 1903 . . . they lived on land given to Mr. Crick by one of his patients for care . . . the property was way back in the woods . . . so primitive that they moved away for 10 years . . . they returned to Pinopolis in 1923 . . . Mrs. Crick . . . would tell it like it was . . . Mr. Crick was more low key . . . They had a chicken farm for 10 years . . . then into the dairy business with three or four cows to milk, delivering to customers in Pinopolis and Moncks Corner . . . In 1947 . . . failing health . . . built a small cottage . . . turned to planting and selling flowers . . . frugal . . . Mrs. Crick made all of her clothes (including underwear) from pretty floral feed sacks . . . their diet was mostly food raised on the farm . . . strawberries, raspberries, figs, pears, vegetables, good rich milk, cream and butter . . . I have seen her sit down and eat a sandwich made with homemade bread, onion, and mayonnaise . . . Some people barely make a track when going down life's highway, others leave deep ruts. Such was the life of the Cricks* (Mildred Owens 1998).

Chicora had a significant influence on the surrounding areas of Moncks Corner, Pinopolis, and Cross for about twenty years. Artifacts from Chicora still appear from time to time. When Robert Austin, who lives near Cross, was going through an old home damaged by Hurricane Hugo in 1989, he found an old trunk containing many postcards with a Chicora postmark. The earliest postmark was 1908. What was unusual was that the postmark was made at the destination, rather than at the origin.

Despite A. S. Emerson's plan for rapid development in New England City, also called Chicora, northerners had not been truly informed about health risks associated with southern swampland. David Tyler remembered: *Malaria fever was so severe until those who didn't die spread out to other places . . . I had a long siege of chronic malaria and lost so much time at school until Papa got Miss Carrie Wiggins to teach me. I remember . . .* [Dr. Richard Y. Dwight] *. . . rolling his pills in a little wooden bowl* [mixing] *arsenic, quinine &* [strychnine]. It is fortunate that Emerson did not prevail in his effort to make Chicora the Berkeley County seat. If malaria had not ended the days of Chicora, Lake Moultrie in 1941 would have flooded it.

In August 1900, the estate of William Cain (1792–1878) sold land to Richard Y. Macbeth (1842–1903) on the northwest quadrant of the intersection of the Berkeley Railroad and the Pinopolis public road (Ball 1983). Macbeth, a Pinopolis businessman from an old St. John's family, built a cotton gin on this

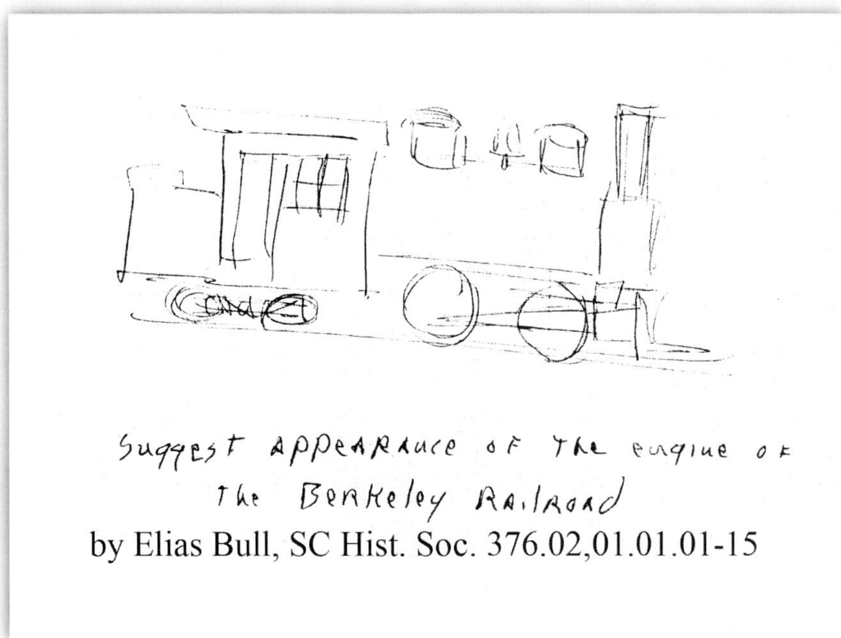

T. P. Ravenel recorded in his diary: "May 11, 1893, Engine on Berkeley RR run up to Pinopolis." This sketch by Elias Bull shows how the engine for the Berkeley Railroad might have looked. *Courtesy of the Elias Bull Collection, South Carolina Historical Society.*

Richard Macbeth built the Pinopolis Cotton Gin in 1900. It remained in operation until 1923. *Courtesy of the Cain files, South Carolina Historical Society.*

property (map:Ⓔ) in October 1900. J. H. Porcher's diary confirms in October 1900: *Richard Macbeth's new gin at RR crossing commenced work today.* Anne S. Ball noted that at Richard Y. Macbeth's death, his widow sold the land *together with the engine, boiler, cotton gin, corn mill, shafting pulleys, belting and all tools in the gin house* to Pinopolis farmers and businessmen William Cain and Henry R. Dwight in May 1906. John Henry Porcher recorded on August 21, 1907: *Henry Dwight and Wm. Cain putting new gin—60 saws & press.*

Farmers from the surrounding areas brought in cotton to be ginned as well as corn to be ground. The chug of the steam engine became a common sound in the village. Its whistle signaled the beginning of the workday. Edgar Davis of Pinopolis, grandson of Henry R. Dwight Sr., still has the steam engine whistle from his grandfather's gin (interview by author, 2005). Henry R. Dwight Jr. recalls that his father's gin had two gongs that were to be struck in case of an emergency. If the gong sounded, a worker would immediately open the steam relief valves and stop the machinery. The gin and mill were closed in 1923. In 1929, the gin property and equipment were sold to Daisy Elnita Harvey (1871–1965), who, at her death, left the property to her nieces, Connie Harvey Murray, Marie Harvey, Elnita Harvey Jennings, and Essie Kizer Harvey (W. Ball 1983).

By the time I was eight years old, my friends and I were allowed to wander anywhere in Pinopolis. The remnants of the old cotton gin building provided a favorite place to visit. One friend, Robert Stevens, remembers some experiences there: *The boiler and some machinery were still in the building around 1944 when I was eight. Kitty and Marianne Davis, Hugh and Harry Burbage, Jimmy Skinner, and*

I would play in the old building. There was a wooden trough, or chute, lined with sheet metal, running from the third floor down to the first floor. The bales of cotton on the third floor must have slid down the chute to the first floor. We kids used the chute as a sliding board. At one place on the second floor, there was a drop of about two feet between the chutes. It was super dangerous to slide. Sure enough, somebody got hurt, and the village fathers tore down the chute. The old building was demolished a few years later (Robert Stevens interview with author, July 22, 2005).

This geodetic chart from 1921 shows the location of the Tram Road.

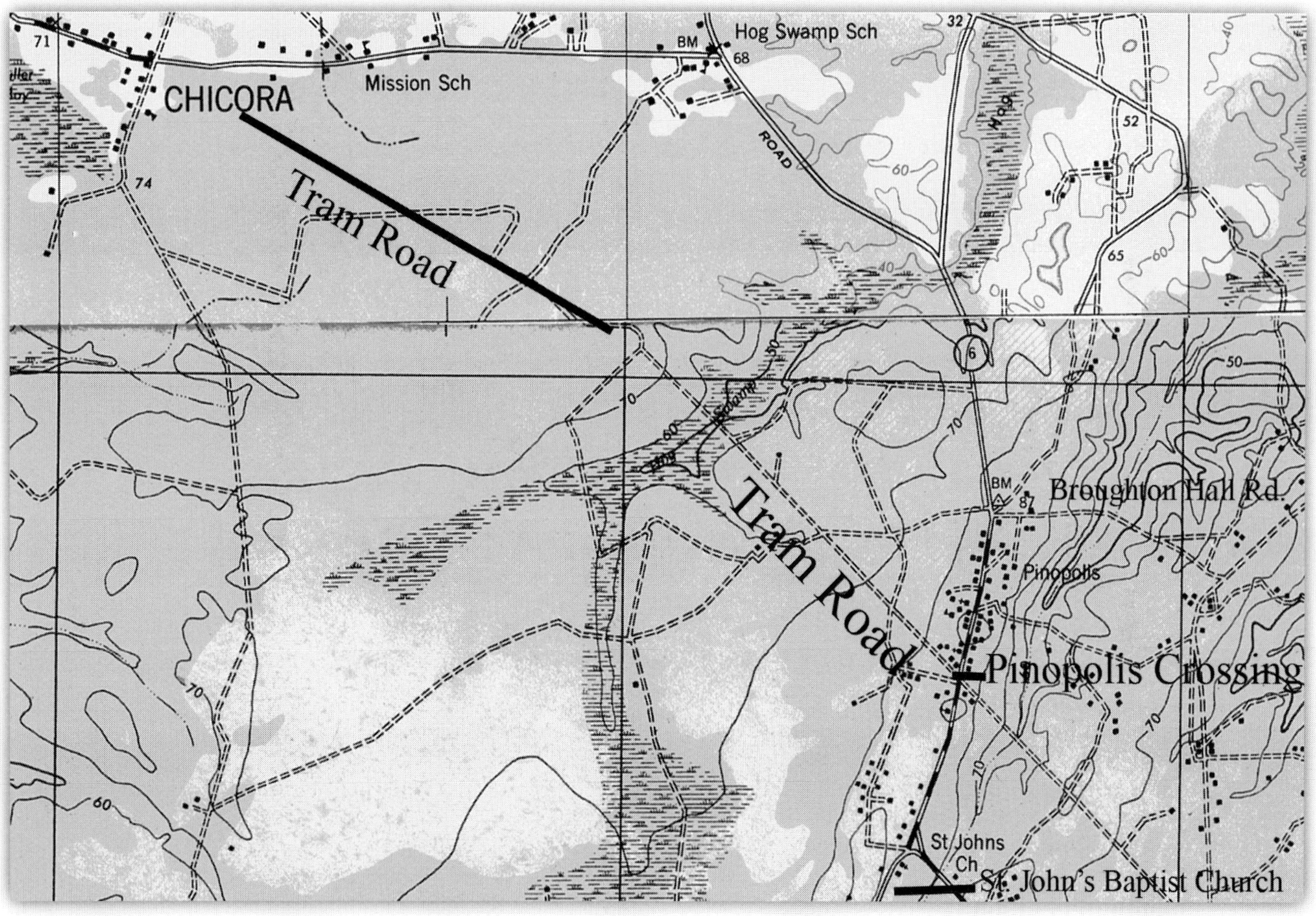

Appendices

Appendix A

People Living in Pinopolis from 1895 to 1911

Handwritten list by Emma Jenkins Gribbin, daughter of Hawkins King Jenkins, in December 1968, Elias Ball Bull papers, Box 8, South Carolina Historical Society

Pinopolis from 1895–1911 as we remember it, and the families who were there during these years. Recorded below as written:

Children of Dr. and Mrs. Joseph P. Cain
1. Henrietta Gourdin 2. William 3. Katherine Macbeth 4. Elizabeth Macbeth 5. Joseph Palmer Jr. 6. Mary Macbeth 7. Julia Anna 8. Charles Macbeth 9. Caroline Porcher 10. James Ravenel 11. Robert Malcolm 12. Francis Gendron 13. Melville Johnson.

Children of Mr. and Mrs. Rene Ravenel
1. Catherine Stevens 2. Isabelle Marion 3. Charles Jervey 4. Rene Jr. 5. James Jervey 6. Stephen Deveaux 7. Maria Jervey 8. Florence Ward 9. Damaris de St. Julien 10. William Jervey

Living with Miss Mary Porcher
Caroline Porcher Heyward; Mr. Mazyck Porcher lived there for about 2 years

Children of Mr. and Mrs. Richard Greene Causey—Jennie M., Sarah E., Richard G. Jr., Mary C., Caroline M., Hugh W.

Children of Mr. and Mrs. Hugh F. Oliver (Mr. Oliver was the Baptist minister at the nearby Church)—Edith, Louise, Hugh F. Jr., Thaddeus, David, Estha Marion called Rose

Living with Miss Lou Harvey
J. P. [Harvey], Stevens [Harvey], Vivian [Harvey]

Son of Mr. Joe Harvey—Harold

Children of Mr. and Mrs. Thomas [West] Williams—Russell, Maude

Children of Mr. and Mrs. Middleton Harvey—Essie, Joseph, Middleton

Children of Mr. and Mrs. Lewis S. Lucas—1.Elizabeth Ravenel [married William Cain] 2. Florence LeNoble 3. Henry Ravenel

Children of Mr. and Mrs. T. P. Ravenel, Jr. (Savannah) [who] lived with Mr. and Mrs. Thomas P. Ravenel [Sr.] and Miss Bessie Ravenel—1. Emily Guerard 2. Elizabeth de St. Julien 3. Mary Wallace [married Thomas Gignilliat] Also Mr. Heyward Ravenel's daughter whose mother had died—Marguerite Hamilton [who married Percy R. Porcher]

Children of Mr. and Mrs. Hawkins King Jenkins—1. Emma Manigault [author of this index] 2. Joseph Evans 3. Hawkins King Jr. 4. Gabriel Manigault

Summer visitors [of] Mr. William Ravenel and "Miss Annie" Ravenel
"The Washington" Ravenels, Julie, Gaillard and William

At Mrs. Maria Porcher's [widow of P. R. Porcher I]—Mr. Isaac Porcher, Mr. John Henry Porcher

Children of Mrs. Rene Ravenel Rene, Jr., Stephen, Maria (married Henry Gaillard), Nina, Percival Porcher Mrs. Ravenel married second time Charles Macbeth

Children of Mr. and Mrs. Richard L. Morrison [sheriff] Leila, Richard, James, Agnes, Lucille, Wells

Children of Mr. and Mrs. S. Porcher Stoney 1. Randall Croft 2. Charlotte Matilda 3. Samuel Porcher, Jr. 4. Laurence O'Hear 5. Thomas Porcher 6. Sanford Barker

Children of Mr. and Mrs. Charles Stevens 1. Elizabeth Allen 2. Laura Ann 3. Henry L. 4. Caroline Sydney (died young)

Children of Mr. and Mrs. Percival Porcher 1. Eudora Mitchell 2. Percival Ravenel Porcher Jr. 3. Richard Dwight

Son of Dr. and Mrs. Richard Y. Dwight—Henry Ravenel

Children of Mrs. William Stoney [in the Massalon house] Kate, Madeleine Gendron, William Edmund

[Additional families that were listed in an index that gave the families already mentioned are:] Mr. and Mrs. Peterson,

Mr. and Mrs. Patrick, Mr. and Mrs. William [Henry] Cain, Mr. and Mrs. Harold Harvey, Mr. and Mrs. Harley [Harlock Webb] Harvey, Mrs. Charles Macbeth, Mrs. Richard Y. Macbeth, Mr. & Mrs. Charles Wiggins

[Mrs. Gribbin also had a page titled] "Old People of the Village"

Miss Anne Stoney and Mr. Dwight Stoney lived with the S. P. Stoneys
Miss Julia and Miss Mary Croft with the Stoneys
Miss Anna Maria Cain and Mrs. Kate Porcher lived with Dr. & Mrs. [Joseph Palmer] Cain
Miss Ellen Gaillard with her sister, Mrs. R.Y. Macbeth
Mr. Stephen DeVeaux with Miss Bessie Ravenel
Mr. Calhoun Cain with Dr. and Mrs. [Joseph Palmer] Cain
Miss Mary Anna Porcher with the Lucas
Miss Sue Jervey with Mr. and Mrs. Rene Ravenel
Miss Mary Couturier lived at the Ophir Lot [with Mrs. Henry F. Porcher]

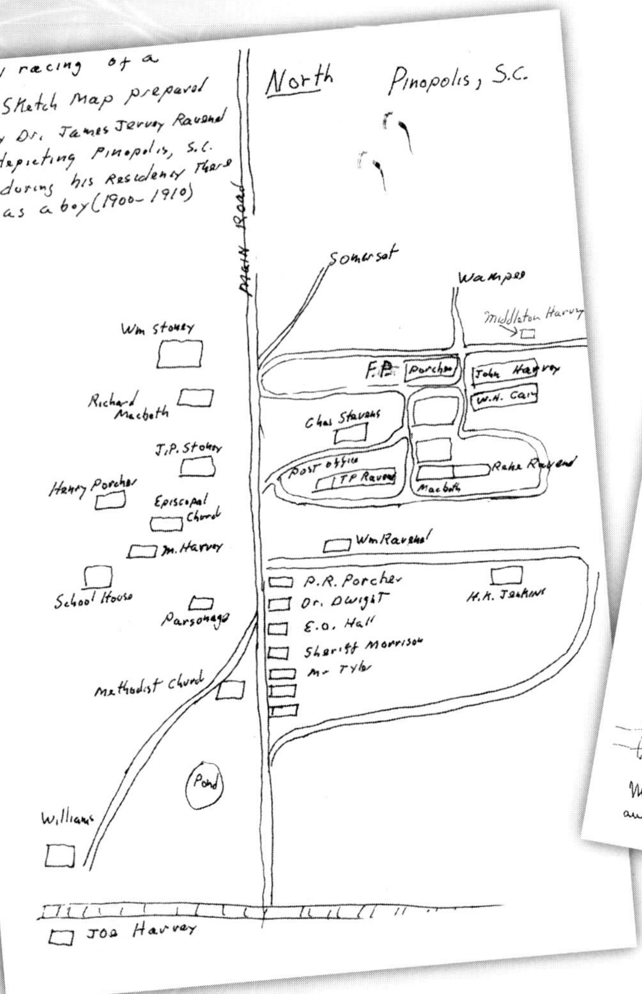

Modification of a map drawn by Dr. James Jervey Ravenel showing Pinopolis as he remembered its layout around 1910. *Courtesy of Martha Porcher Ravenel.*

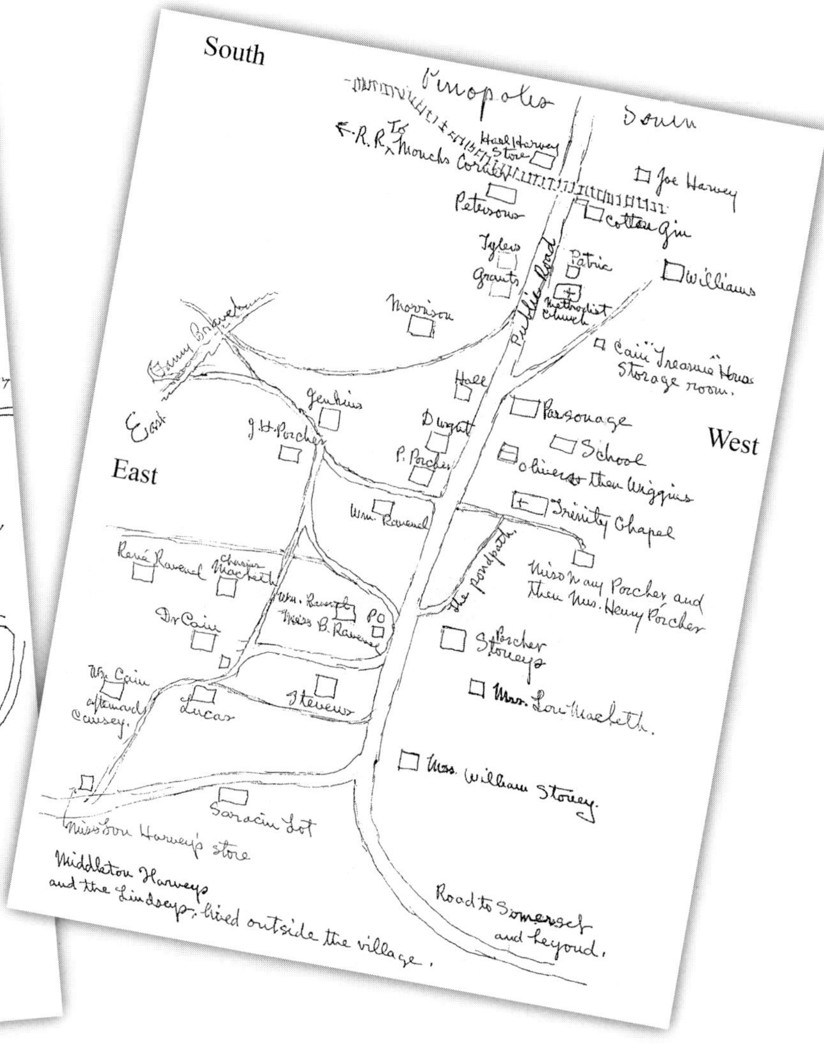

Map drawn by Emma Gribbin showing Pinopolis as she remembered it from 1895 to 1911. *Courtesy of Elias Bull Collection, South Carolina Historical Society.*

Appendix B

Teachers and Students of the Pinopolis Academy and Public School

It was the custom in the minutes to give a teacher's name and previous residence. The following list of teachers at the school is from Ball with some notes I made from the minutes.

Virgil C. Dibble, first principal, July 1, 1896

Other principals were Harold Langley; T. C. Hannah; Mr. Allan; Macmilliam Seabrook; Mrs. Annie Adams of Virginia; Miss Tobin; and Mr. Wyndham Manning of Pinopolis.

Some of the teachers were Miss Lou Walpole and Miss Martha Rivers of Johns Island; and Miss Eunice Horton of Aiken. From Pinopolis were Miss Caroline Cain; Mrs. Joseph Edwards (mother of Margaret Stevens); and Miss Moselle Skinner (sister of Harriet Vardell and Julia Skinner). Others were Miss Julia Ball; Miss Virgie C. Dibble; Mr. O'Driscol; Miss Edna Tobin; Miss Kate Tobin; Mr. Deveaux Jervey; Mr. William McCleod; Mr. Mac Mazyck; Miss Marion McCoy; Annie Neil; Noni Simons (Mrs. Sears Pegues); Ernestine Chestnut Stoney; Miss Essie Harvey; Miss Ward; and Miss Rose Ellis, who became a principal. There were three teachers named Hall: Miss C. P. Hall, Miss Annie Hall (assistant principal in 1899 who resigned in 1901), and Miss Katie Hall who succeeded her.

Students 1922–23

Fourth Grade—Vivian Clarke, Adelaide Dennis (Dick Lacey), Stevie Rudloff, and Robert (no last name given)

Fifth Grade—Joseph (no last name), Sears Brewer, Mamie (no last name), Eunice Hare, and Pledger Parker

Sixth Grade—Hallie Scarborough

Seventh Grade—Charles Cain, Thelma Clarke, Beatrice Dennis (Bea Plyler), Henry Dwight, Emily Fishburne, John Harvey, and May Youngblood

Students 1923–24

Seventh Grade—John Harvey and Hallie Scarborough

Eighth Grade—Charles Cain, Henry Dwight, Emily Fishburne, Allen Scarborough, and Essie Harvey

Appendix C

Partial Family Genealogies

Descendants of William Cain

1 William Cain 1792–1878
.. +Anne Porcher Palmer 1804–1855
.... 2 William Henry Cain 1834–1909
.... +Henrietta Macbeth 1837–1903
.... 3 Charles Macbeth Cain 1859–
.... 3 Ann Palmer Cain 1860–1931
.... 3 Melville Cain 1865–1947
.... +John Irving Westervelt Jr.
.... 2 John Calhoun Cain 1835–1916
.... +Charlotte Horlbeck
.... 3 Harriet Chisholm Cain 1869–1950
.... +Robert Wilson II, MD 1867–1946
.... 4 Robert Wilson Jr. 1905–1985
.... 4 Harriet C. Wilson 1908–
.... 4 James M. Wilson 1911–1982
.... 2 Dr. Joseph Palmer Cain 1836–1903
.... +Mary Catherine Macbeth 1851–1926
.... 3 William Cain 1874–1926
.... +Elizabeth Ravenel Lucas 1883–1941
.... 4 William Cain Jr. 1908–1993
.... +Mary Jane Patrick 1909–1970
.... 5 William Cain 1947–
.... 5 Sarah Patrick Cain 1948–
.... +James Spruill
.... 4 Charles Lucas Cain 1910–1981
.... 4 Joseph Palmer Cain 1914–
.... +Bertha Smith
.... 4 Elizabeth Ravenel Cain 1914–1976
.... +Lemuel Whitaker Boykin III
.... 5 Elizabeth Ravenel Boykin 1943–
.... 5 Lemuel Whitaker Boykin IV 1946–
.... 5 Marcia Stockton Boykin 1950–
.... 3 Katherine Macbeth Cain 1875–1956
.... 3 Elizabeth Macbeth Cain 1877–1930
.... +Henry Ravenel Dwight 1873–1961
.... 4 Mary Macbeth Dwight 1905–1983
.... +William Edgar Davis
.... 5 Katherine Cain Davis 1933–
.... +Bratton de Loach
.... 5 Marianne Gillican Davis
.... +Don Monroe

William Cain

Dr. Joseph Palmer Cain

Elizabeth Macbeth Cain

Henry Ravenel Dwight Sr.

.... 5 William Edgar Davis Jr.
.... 4 Rowena Ravenel Dwight 1909–1991
.... +John Emerson Langley –1964
.... 5 Elizabeth Dwight Langley 1934–
.... 5 Patricia Dwight Langley 1939–
.... 5 Anne Ravenel Langley 1949–
.... 4 Henry Ravenel Dwight Jr. 1911–
.... 3 Caroline Porcher Cain 1883–1958
.... 3 Robert Malcolm Cain 1888–1970
.... 3 Dr. Francis Gendron Cain 1889–1977
.... +Lena Parham Atkins
.... 4 Francis Gendron Cain
.... 4 James Ravenel Cain 1925–
.... 4 William Hubbard Cain 1928–
.... 3 James Ravenel Cain 1885–1970
.... +Isabelle Lindsay
.... 2 Anna Maria Cain 1841–1905
.... 2 Edward Greenland Cain 1845–
.... +Emily Ravenel
.... 3 Elizabeth Snowden Cain 1870–1947
.... +Michael Kenan
.... 4 Evelyn Elizabeth Kenan 1895–
.... +Lovell L. Oliveros
.... 5 Robert Lovell Oliveros 1920–
.... +Rosa Richards Burnet

Descendants of Brutus Green

1 Brutus Green 1856–
.. +Satara 1860–
.... 2 Charlie Green 1892–
.... 2 Annie Green 1894–
.... +Elias Brown
.... 3 Franklin Brown –1983
.... +Lucille Rachel Grant –1994
.... 4 Louann Brown 1939–
.... 3 George Brown
.... +Elsie Washington
.... 4 Albertha Brown 1949–
.... +Mr. Varner
.... 4 Idethia Brown
.... 4 Annie Mae Brown
.... 4 Dorothy Brown
.... 4 Joan Brown
.... 4 Clarisa Brown
.... 4 Janie Dean Brown
.... 4 Madilyn Brown

.... 3 Walter Brown
.... 3 Elias Brown Jr.
.... 3 Andrew Brown
.... 3 Elizabeth Brown
.... +Isaac Wilson
.... 3 Thomas Brown
.... +Isabella Burch
.... 2 Thomas Green 1896–
.... 2 Daniel Green 1900–
.... 2 Ida Green 1902–
.... 2 Sallie Green 1904–
.... 2 Rosa Green 1906–

Descendants of Nathan Hall Guyton

1 Nathan Hall Guyton 1818–1890
.. +Elizabeth Harriet Lynes 1835–1890
.... 2 Walter G. Guyton 1856–1929
.... +Annie Viola Sessions 1887–
.... 3 Louise Guyton
.... +Samuel Pledger Parker 1885–1917
.... 4 Anne Louise Parker 1910–
.... +Burbage
.... 4 Samuel Pledger Parker Jr. 1912–
.... +Millie
.... 5 Samuel Pledger Parker III 1943–
.... 5 Suzanne Parker 1944–
.... 4 Elizabeth Parker 1914–
.... +Marion
.... 4 Julia Parker 1916–
.... +Thomas Young 1904–1989
.... 5 Judy Young 1942–
.... 5 Thomas Young 1946–
.... 2 Julia Varina Guyton 1861–1937
.... 2 George N. Guyton 1866–1927
.... 2 Margaret Annie Guyton 1868–1950
.... +James Bloodworth Skinner 1851–1900
.... 3 Moselle Skinner 1895–1994
.... 3 James Bloodworth Skinner 1897–1939
.... +Lillian
.... 4 Mary Anne Skinner
.... 4 James Bloodworth Skinner
.... 4 Thomas Skinner
.... 4 William Skinner
.... 3 Harriet Skinner 1899–1985
.... +James Cain Vardell Jr. 1897–1957
.... 3 Julia Skinner 1901–1993

.... 2 Horace Waldo Guyton 1870–1942
.... +Margaret W. 1876–
.... 2 Sidney DeMartin Guyton 1873–1953
.... +Eva Paula McCants
.... 2 Rev. Arthur Fitzgerald Guyton 1876–1954
.... +Margaret Francis Hare
.... 3 Margaret Elizabeth Guyton
.... +X Gartelmann
.... 4 Margaret Gartelmann
.... +X Waters

Joseph Arnold Harvey

Descendants of Arnold Harvey

1 Arnold Harvey 1754–1834
.. +Cynthia Wood 1761–1812
.... 2 John Harvey 1778–1823
.... +Jemima Huxford 1784–1841
.... 3 Arnold James Harvey 1804–1872
.... +Belisant McCants 1810–
.... 4 Julia Arnold Harvey 1833–1862
.... +Joseph Morton Clarke –1892
.... 3 Harlock Huxford Harvey 1806–1883
.... +Margaret Francis Evans 1825–1904
.... 4 Joseph Arnold Harvey 1851–1937
.... +Eva Eugenia Bobo 1854–1905
.... 5 Harold Huxford Harvey Sr. 1888–1960
.... +Annie Lucille McCants 1894–
.... 6 Julia Evelyn Harvey 1918–
.... +Raymond A. Weathers 1920–
.... 7 Raymond Harvey Weathers 1946–
.... 7 Donald Edwards Weathers 1952–
.... +Amber Metts 1956–
.... 6 Harold Huxford Harvey 1920–
.... +Francis Joseph Arnold 1924–
.... 7 Linda Joe Harvey 1949–
.... 7 Harold Huxford Harvey III 1951–
.... +Naomi
.... 7 Robert Marvin Harvey 1955–
.... 4 [2] Catherine Almira Harvey 1853–1922
.... +[1] Thomas Middleton Harvey 1842–1892
.... 5 [3] Lillian Ester Harvey 1880–1956
.... +[4] Herbert Jeter Thomas –1953
.... 5 [5] Thomas Middleton Harvey 1884–1949
.... 5 [6] Joseph Harlock Harvey 1887–1947
.... +[7] Evelyn deMedicis
.... 4 Harlock Webb Harvey 1867–1930
.... +Nettie Camilla Thomas 1869–1915

Harlock Webb Harvey

.... 5 Harlock Walter Harvey 1892–
.... 5 Nettie Elizabeth Harvey 1900–1979
.... +Ernest Edgar Bishop 1897–1972
.... 6 Nettie Camilla Bishop
.... 5 Reuben Thomas Harvey 1903–1962
.... +Agnes Elizabeth Parker 1913–
.... 6 Florence Agnes Harvey 1936–
.... 6 Nina Bessie Harvey 1938–
.... 6 Thomas Anne Harvey 1943–
.... 6 Ronald Theodore Harvey 1957–
.... 3 Thomas Joseph Harvey 1815–1880
.... +Jane Margaret Purkey 1825–1888
.... 4 [1] Thomas Middleton Harvey 1842–1892
.... +[2] Catharine Almira Harvey 1853–1922
.... 5 [3] Lillian Ester Harvey 1880–1956
.... +[4] Herbert Jeter Thomas –1953
.... 5 [5] Thomas Middleton Harvey 1884–1949
.... 5 [6] Joseph Harlock Harvey 1887–1947
.... +[7] Evelyn deMedicis
.... 4 Eudora Harvey 1845–1907
.... +Robert R. Lindsay 1843–1905
.... 5 Bertha Maude Lindsay 1876–
.... 5 Martha Jane Lindsay 1880–1913
.... +Charles Ashley Harvey 1862–1931
.... 6 Thomas Lindsay Harvey 1908–
.... 6 Robert Marion Harvey 1913–
.... 5 Joseph Harvey Lindsay 1883–1963
.... 4 Louise Octavia Harvey 1848–1910
.... 4 John Purkey Harvey 1851–1897
.... +Ann Chandler Stevens 1856–1892
.... 5 Vivian Jane Harvey 1885–1954
.... +William Leander Wiley 1883–
.... 6 Estelle Wiley 1910–
.... 6 Stevens Harvey Wiley 1913–
.... +Tula Baker 1913–
.... 6 Cecil Wiley 1919–
.... +Dorothy Arnold
.... 6 William Wiley 1922–
.... +Dora Jean Purvis 1922–
.... 5 Stevens McTeer Harvey
.... 4 Sara Eliza Harvey 1854–1936
.... 2 Thomas Harvey 1784–1836
.... +Mary Huxford 1795–1837
.... 3 John Arnold Harvey 1813–1862
.... +Catherine B. Smith 1818–1866
.... 4 William Henry Harvey 1842–1875

Nettie Elizabeth Harvey

Thomas Middleton Harvey

Catharine Almira Harvey

J. Goodlow Stevens

.... +Dorcas Elizabeth Crawford 1848–1888
.... 5 William Mortimer Harvey 1866–1938
.... +Sallie Estelle Godfrey 1870–1952
.... 6 Gladys Harvey 1894–
.... 6 Myrtle Harvey 1895–
.... 6 Alva Vernon Harvey 1897–
.... 6 Clemie Harvey 1900–
.... 6 Virgil Harvey 1900–
.... 6 Glaser Ferraire Harvey 1903–
.... 6 Ashley Vance Harvey 1904–
.... 6 William Mortimer Harvey 1906–
.... 6 John Kershaw Harvey 1910–
.... 5 Lewis Henry Harvey 1869–1924
.... +Minnie Elizabeth Beatson 1868–
.... 6 Mollie Harvey 1900–1911
.... 6 Connie Harvey 1901–
.... +Jessie Murray 1894–1955
.... 7 Velma Connie Murray 1926–
.... 6 Marie Harvey 1903–
.... 6 Elnita Harvey 1904–
.... 6 Lewis Harvey Jr. 1906–
.... 6 Essie Harvey 1909–
.... +W.P. Kizer
.... 7 Gwendolyn Essie Kizer 1936–
.... +Wilson Reeves Lewis
.... 7 Patricia Marie Kizer
.... 5 Daisy Elnita Harvey 1871–1965

Lewis Henry Harvey

Minnie Elizabeth Beatson

Descendants of James Macbeth

1 Charles Johnston 1732–1804
.. +Mary McKenzie 1753–1796
.... 2 Catherine Johnston 1771–1845
.... +James Macbeth 1767–1822
.... 3 Mary Macbeth 1800–1865
.... +William Robertson 1796–
.... 3 Hon. Charles Macbeth 1805–1881
.... +Henrietta Gourdin Ravenel 1816–1888
.... 4 Charles Johston Macbeth 1835–
.... +Isabelle M. De Veaux
.... 4 Henrietta Macbeth 1837–1905
.... +William Henry Cain
.... 5 Charles Macbeth Cain
.... 5 Ann Palmer Cain
.... 5 Henry Stevens Cain
.... 5 Melville Cain
.... 5 James Ravenel Cain

.... 4 James Ravenel Macbeth
.... 4 Richard Yeadon Macbeth 1842–1903
.... +Mary Louisa Gaillard 1851–1914
.... 4 Mary Catherine Macbeth 1851–1926
.... +Dr. Joseph P. Cain 1836–1903
.... 5 Dr. Joseph Palmer Cain 1836–1903
.... 6 William Cain 1874–1926
.... +Elizabeth Ravenel Lucas 1883–1941
.... 5 William Henry Cain
.... 5 John Calhoun Cain
.... 5 Edward Greenland Cain

Descendants of Michael Major

1 Michael Major
.. +Evelenna Palmer
.... 2 Gallager Major
.... 3 Barbara Hampton Major
.... 2 Sally Major
.... +Noah Frazier
.... 3 Herman Frazier
.... 3 Ernest Frazier
.... 3 Essie Frazier
.... 2 Andrew Major
.... +Mary Pou
.... 3 Ethel Pou Major
.... 2 Dirinda Major –1952
.... +Richardson Smalls –1935
.... 3 Rosa Bell Smalls
.... +Henry Coaxum
.... 3 Maggie Smalls
.... +Jacob Vanish
.... 3 Edward Lute Smalls
.... +Rebecca Bryant
.... 4 William Smalls 1933–
.... +Justine England 1935–
.... 3 Sarah Ding Smalls
.... +Theodore Howell
.... 3 Eunice Rebecca Smalls
.... 3 Julia Smalls
.... +Elijah White
.... 3 Elizabeth Smalls
.... +Rev. Elijah McCants

Descendants of McCants

1 McCants
.... 2 Louvinia McCants

.... +Broughton A. Gibbs
.... 3 Anna Gibbs Brown
.... 4 Frank Brown
.... 3 Minnie Elizabeth Gibbs
.... +Eugene Heyward
.... 4 Elise Heyward
.... +Haynes
.... 4 Aleathea Heyward
.... 4 Samuel Heyward 1928–
.... 4 Mary A. Heyward
.... +Singleton
.... 4 Alvin E. Heyward
.... 4 Freddie Heyward
.... 4 Gladys Heyward
.... 3 Rebecca Gibbs
.... +Mack
.... 4 Leroy Mack
.... +Lillian
.... 4 Gertie Mack
.... +Ernest Smith
.... 4 Walter Ervin Mack
.... +Mildret
.... *2nd Husband of Rebecca Gibbs:
.... +Alfred Major
.... 3 Willie Gibbs
.... +Rosalee
.... 3 Alex Gibbs
.... 3 Elia Gibbs
.... +Richard Wright
.... 4 Richard Wright Jr.
.... 5 Bitty
.... 2 Jeulla McCants
.... +Peter Green
.... *2nd Husband of Jeulla McCants:
.... +Name Gadsden
.... 3 Martha Gadsden
.... 3 Courtney Gadsden
.... 3 Fred Gadsden
.... 3 Milton Gadsden
.... 3 Earnest Gadsden
.... 3 Martha Gadsden
.... 3 Dora Gadsden
.... 2 Mazine McCants
.... 3 Annie McCants
.... 2 Elvidra McCants
.... 3 Mattie McCants

.... 3 Joseph McCants
.... 3 Matthew McCants
.... 3 Sidney McCants
.... 3 Ida McCants
.... 3 Effie McCants

Descendants of Lena Perry

1 Lena Perry 1825–
.... 2 Nathan Perry 1869–
.... +Florence 1867–
.... 3 Malcolm Perry 1890–
.... 3 Mary Eliza Perry 1893–1974
.... +Robert Palmer
.... 4 Francis Palmer –2000
.... 4 William Perry
.... +Rosalee Ferguson 1915–
.... 5 Cynthia Perry
.... +X Bryant
.... 5 Dianna Perry
.... +Y Savage
.... 4 Emily Palmer 1933–
.... 5 Maxine Palmer
.... 5 Roberta Palmer
.... 4 Catherine Palmer –1951
.... 4 Emmatrude Palmer –1972
.... 4 Thaddeus Perry Palmer 1937–
.... 3 Catherine Perry 1894–1994
.... +Edward Beatty 1875–1953
.... 4 Wilhelmina Beatty
.... +X Morris
.... 5 Willa Mae Laverne B. Mazyck 1942–
.... 5 Ophelia Beatty Morris 1944–
.... 5 Sandra J. Beatty Morris 1944–
.... 5 William James Beatty Morris 1951–
.... +Sharon Heyward
.... 3 Bessie Perry 1898–
.... 3 Evelyn Perry
.... +Howard Shine
.... 4 Adrian Shine 1930–

Descendants of Isaac Porcher I

1 Isaac Porcher I –1727
.. +Claude Cherigny 1661–1726
.... 2 Isaac Porcher II 1682–1743
.... 2 Peter Porcher I 1695–1753
.... +Charlotte Marianne Gendron 1695–1729

.... 3 Peter Porcher II 1726–1781

.... +Elizabeth Cordes 1727–1783

.... 4 Peter Porcher III 1756–1793

.... +Elizabeth Marion 1760–1796

.... 5 Elizabeth Catherine Porcher 1781–1841

.... +Joseph Palmer 1776–1841

.... 6 Anne Palmer 1804–1855

.... +William Cain 1792–1878

.... 4 Elizabeth Porcher 1760–1804

.... +Peter Gaillard 1757–1833

.... 5 Peter Gaillard 1783–1843

.... +Elizabeth Gourdin 1786–1836

.... 6 Harriet Porcher Gaillard 1826–1893

.... +Samuel Gaillard Stoney 1853–1922

.... 4 Col. Thomas Porcher 1766–1836

.... +Elizabeth Sinkler DuBose 1788–1866

.... 5 Martha Maria Porcher 1807–1842

.... +Isaac Marion Dwight 1799–1873

.... 6 Richard Yeadon Dwight

.... +Rowena Elizabeth Ravenel

.... 7 Henry Ravenel Dwight 1873–1961

.... 6 Eudora W. Dwight 1835–1904

.... +Francis Marion Mitchell

.... 7 Harriet Marion Mitchell 1867–1957

.... +Richard Singleton Kirk 1866–1928

.... 8 Gabrielle Marion Kirk

.... +William Snowden Gaillard

.... 8 Francis Marion Kirk

.... +Videau Marion Legare Beckwith

.... 8 Eudora Dwight Kirk

.... +Frank McLeod

.... 8 Richard Singleton Kirk

.... +Martha Kirk

.... 5 Anna Maria Porcher 1811–

.... +Peter Gaillard Stoney 1809–1884

.... 6 Samuel Porcher Stoney

.... 5 Peter Cordes Porcher 1814–1853

.... 5 Henry Francis Porcher 1819–1860

.... +Caroline Lodieske Couturier

.... 6 Henry Francis Porcher 1854–1903

.... +Margaret Deas Ravenel

.... *2nd Wife of Col. Thomas Porcher:

.... +Charlotte Mazyck –1804

.... 5 Thomas Porcher 1796–1843

.... +Catherine Gaillard 1795–1874

.... 5 Elizabeth Catherine Porcher 1798–1846

.... +Henry Ravenel
.... 6 Thomas Porcher Ravenel 1824–
.... +Elizabeth Margaret Wilson 1827–
.... 7 Samuel Wilson Ravenel
.... +Margaret Kinney
.... 7 Thomas Porcher Ravenel 1850–
.... +Emily Howard
.... 7 Elizabeth Catherine Porcher Ravenel 1852–
.... 7 Margaret Deas Ravenel
.... 7 Florence Hartley Ravenel 1860–
.... +Lewis S. Lucas
.... 7 Heyward Ravenel 1868–
.... +Marguerite Hamilton Seabrook
.... 8 [2] Marguerite Hamilton Ravenel 1898–1965
.... +[1] Percival Ravenel Porcher III 1897–1965
.... 9 [3] Percival Ravenel Porcher IV 1931–
.... 9 [4] Margaret Deas Porcher 1933–
.... 9 [5] Heyward Ravenel Porcher 1940–
.... 6 Rene Ravenel 1826–1875
.... +Isabelle Deveaux
.... 7 Rene Ravenel 1856–1901
.... 6 [7] Maria Catherine Ravenel 1832–1907
.... +[6] Percival Ravenel Porcher I 1829–1864
.... 7 [8] Isaac de Cherigny Porcher 1859–1933
.... 7 [9] Percival Ravenel Porcher II 1861–1931
.... +[10] Martha Dwight Mitchell 1858–1932
.... 8 [1] Percival Ravenel Porcher III 1897–1965
.... +[2] Marguerite Hamilton Ravenel 1898–1965
.... 9 [3] Percival Ravenel Porcher IV 1931–
.... 9 [4] Margaret Deas Porcher 1933–
.... 9 [5] Heyward Ravenel Porcher 1940–
.... 8 [11] Richard Dwight Porcher 1900–1973
.... +[12] Miriam Ellison 1906–1975
.... 9 [13] Martha Ravenel Porcher 1935–
.... 9 [14] Richard Dwight Porcher Jr. 1939–
.... 9 [15] Henry Ravenel Porcher 1940–
.... 6 Rowena Elizabeth Ravenel 1837–1924
.... 5 Dr. William Porcher 1800–1833
.... +Isabella Sarah Peyre 1803–1890
.... 6 Dr. Francis Peyre Porcher 1824–1895
.... +Virginia Leigh 1835–1866
.... 7 Dr. Walter Peyre Porcher 1858–1912
.... 7 Virginia Leigh Porcher 1866–1940
.... *2nd Wife of Dr. Francis Peyre Porcher:
.... +Margaret Bently Ward 1849–1895
.... 7 Anne Allston Porcher 1882–1970

Marguerite Hamilton Seabrook

Marguerite Hamilton Ravenel

Percival Ravenel Porcher III

Dr. Francis Peyre Porcher

.... 6 Julian Henry Porcher 1828–1908
.... 4 Samuel Porcher 1768–1851
.... +Harriet Porcher 1772–1843
.... 5 Harriet Porcher 1792–1825
.... +James Gaillard 1788–
.... 6 Peter Charles Gaillard 1812–1889
.... +Anne Lawrence Snowden 1819–1871
.... 7 Cleremonde Serre Gaillard 1843–1914
.... +William Henry Sinkler I
.... 5 Thomas William Porcher 1807–1889
.... +Elinor Cordes Gaillard 1810–1888
.... 6 Julius Theodore Porcher 1829–1863
.... +Mary Fanning Wickham
.... 7 Anne Wickham Porcher
.... +Charles St. George Sinkler
.... 3 Philip Porcher I 1730–1801
.... +Mary Mazyck 1734–1785
.... 4 George Porcher 1775–1831
.... +Marianne Gendron Palmer 1784–1835
.... 5 Frederick Adolphus Porcher 1809–1888
.... +Rebecca Branford Rhodes
.... 5 Dr. John Palmer Porcher 1811–1839
.... +Anna Ford Banks 1818–1844
.... 6 John Palmer Porcher Jr. 1836–1864
.... +Catherine Cordes Deveaux –1923
.... 7 George Porcher 1863–1869
.... 4 Isaac Porcher III 1778–1849
.... +Mary Jarret Weston 1781–1808
.... 5 Isaac Porcher 1803–1809
.... *2nd Wife of Isaac Porcher III:
.... +Mary Stevens 1793–1824
.... 5 Isaac Porcher V 1812–1849
.... +Elizabeth L. Gaillard 1814–1870
.... 6 Isaac Porcher VI 1842–1861
.... 6 Phillip Porcher 1850–
.... *3rd Wife of Isaac Porcher III:
.... +Charlotte Mazyck Ravenel 1795–1835
.... 5 [6] Percival Ravenel Porcher I 1829–1864
.... +[7] Maria Catherine Ravenel 1832–1907
.... 6 [8] Isaac de Cherigny Porcher 1859–1933
.... 6 [9] Percival Ravenel Porcher II 1861–1931
.... +[10] Martha Dwight Mitchell 1858–1932
.... 7 Eudora Mitchell Porcher 1894–1903
.... 7 [1] Percival Ravenel Porcher III 1897–1965
.... +[2] Marguerite Hamilton Ravenel 1898–1965
.... 8 [3] Percival Ravenel Porcher IV 1931–
.... 8 [4] Margaret Deas Porcher 1933–

PR Porcher I and Maria Ravenel

PR Porcher II

Martha Dwight Mitchell

Eudora Mitchell Porcher

Isaac de Cherigny Porcher

.... 8 [5] Heyward Ravenel Porcher 1940–
.... 7 [11] Richard Dwight Porcher 1900–1973
.... +[12] Miriam Ellison 1906–1975
.... 8 [13] Martha Mitchell Porcher 1935–
.... 8 [14] Richard Dwight Porcher Jr. 1939–
.... 8 [15] Henry Ravenel Porcher 1940–
.... 5 John Henry Porcher 1835–1910

John Henry Porcher

Descendants of Daniel Ravenel

1 Daniel Ravenel 1629–
.... 2 Rene Ravenel 1656–
.... 3 Rene Louis Ravenel 1694–
.... 4 Henry Ravenel 1729–1785
.... 5 Rene Ravenel 1762–1822
.... 6 Dr. Henry Ravenel 1790–1867
.... +Catherine Stevens 1792–1816
.... 7 Henry William Ravenel 1814–1887
.... +Elizabeth G. Snowden
.... 8 Emily Ravenel 1846–1908
.... +Edward G. Cain 1845–
.... *2nd Wife of Dr. Henry Ravenel:
.... +Mary Ester Dwight 1797–1820
.... *3rd Wife of Dr. Henry Ravenel:
.... +Elizabeth Catherine Porcher 1798–1846
.... 7 Thomas Porcher Ravenel 1824–1898
.... +Elizabeth Margaret Wilson 1827–
.... 8 Samuel Wilson Ravenel 1848–1922
.... +Margueritte Kinney
.... 8 Thomas Porcher Ravenel II 1850–1922
.... +Emily Howard
.... 9 Emily Guerard Ravenel 1890–
.... 9 Elizabeth de St. Julien Ravenel 1893–
.... 9 Mary Wallace Ravenel 1898–
.... +Thomas H. Gignilliat
.... 10 Margaret Deas Gignilliat 1931–
.... 10 Mary Ravenel Gignilliat 1936–
.... 10 Thomas Heyward Gignilliat 1938–
.... 8 Elizabeth Catherine P. Ravenel 1852–1935
.... 8 Margaret Deas Ravenel 1857–1944
.... +Henry Francis Porcher
.... 8 Florence Hartley Ravenel 1860–1945
.... +Lewis S. Lucas 1859–1915
.... 9 Elizabeth Ravenel Lucas 1883–1941
.... +William Cain 1874–1926
.... 9 Florence LeNoble Lucas 1889–1977
.... 9 Henry Ravenel Lucas 1891–1973

Thomas Porcher Ravenel and Elizabeth Margaret Wilson

Elizabeth Catherine P. Ravenel

.... 8 Heyward Ravenel 1868–1903
.... +Marguerittе Hamilton Seabrook
.... 9 Margueritte Hamilton Ravenel 1898–1965
.... +Percival Ravenel Porcher 1897–1965
.... 7 Dr. Rene Ravenel 1826–1875
.... +Isabelle M. Deveaux
.... 8 Rene Ravenel 1856–1901
.... +Maria Stevens Jervey 1856–1929
.... 9 Dr. James Jervey Ravenel 1890–1984
.... +Dorothy Wallace
.... 9 Maria Jervey Ravenel 1892–1991
.... 9 Dr. William Jervey Ravenel 1901–1981
.... 7 Maria Catherine Ravenel 1832–
.... 7 Rowena E. Ravenel 1837–
.... +Dr. Richard Yeadon Dwight
.... 8 Henry Ravenel Dwight 1873–1961
.... 7 William Francis Ravenel 1828–1906
.... +Ellen DuBose
.... 8 Anne Stevens Ravenel 1865–1954

Descendants of Merriman Ravenel

1 Merriman Ravenel
.. +Mary Read
.... 2 Eliza Ravenel
.... +Joe Sheppard
.... 3 Mary Rebecca Sheppard 1917–
.... +McKinley Smalls –1938
.... 4 Gertie Smalls 1933–2001
.... 4 Maude Ruth Smalls 1935–
.... 4 Francis Smalls 1938–
.... *2nd Husband of Mary Rebecca Sheppard:
.... +Andrew Brown
.... 4 Andrew James Brown 1946–2000
.... 4 Christopher Brown 1948–

Descendants of Col. Morton Alexander Waring

1 Col. Morton Alexander Waring 1784–1863
.. +Rebecca Hamilton 1786–1871
.... 2 Paul Hamilton Waring 1805–1881
.... +Sarah Wilkinson Glover 1805–1876
.... 2 Rebecca B. Waring 1808–1855
.... 2 Dr. Morton Alexander Waring 1809–1875
.... +Anne Henrietta Rhodes 1813–1896
.... 3 Dr. Morton Nathaniel Waring 1831–1882
.... +Rebecca Hamilton Webb 1835–1920

.... 4 Martha Caroline Waring 1858–1858
.... 4 Morton Alexander Waring 1859–1859
.... 4 Anne Rhodes Waring 1860–1860
.... 4 Eliza Ladson Waring 1861–1926
.... 3 John Rhodes Waring 1832–1901
.... 3 Rebecca Hamilton Waring 1834–1855
.... 3 Susan Edith Waring 1844–1907
.... 3 Mary Rhodes Waring 1848–
.... +Robert Y. H. Hennegan 1848–
.... 2 Susan Mary Waring 1812–1870
.... +Thomas Ladson Webb
.... 2 Mary Hamilton Waring 1820–1885
.... +Thomas Whaley
.... 2 Archibald Hamilton Waring 1834–1875
.... +Hannah Elizabeth Pawley 1836–1879
.... 3 Archibald Hamilton Waring 1856–1929
.... 3 Morton Alexander Waring 1863–1921
.... +Mary Margaret Webb 1872–1946
.... *2nd Wife of Morton Alexander Waring:
.... +Martha Woods Gibson 1872–1899
.... 3 George Pawley Waring 1868–1933

Descendants of W. Capers Williams

1 W. Capers Williams 1834–1862
.. +Francis Elizabeth 1838–1915
..... 2 Thomas West Williams 1857–1922
.... +Carrie Mellard Wiggins 1882–
.... 3 James Russell Williams Sr. 1883–1947
.... +Olive Exum Bardin 1888–1963
.... 4 James Russell Williams Jr. 1909–1973
.... +Clara Lee Peeples 1912–1992
.... 5 LaClaire Williams Laffitte
.... 5 Angela Williams
.... 5 James Russell Williams III 1943–
.... +Sandra Elaine Smith 1945–
.... 6 Brett Russell Williams 1963–
.... 7 Lee Russell Williams 1990–
.... 6 Todd Bardin Williams 1964–
.... 3 Maude Lillian Williams 1885–1966

IN MEMORIAM
JOHN HENRY
PORCHER.
MARCH 9. 1835.
DEC. 28. 1910.
MORTON NATHANIEL
WARING. M.D.
BORN SEPT. 6. 1831
DIED SEPT. 21. 1882.
RENE RAVENEL,
M. D.
NOV. 13TH. 1826.
JULY 27TH. 1875.
"SWEET BE THY REST,
TILL HE BIDS THEE ARISE,
TO HAIL HIM IN TRIUMPH
DESCENDING THE SKY."
H
HARLIE WEBB
HARVEY
OCT. 28, 1867
MAR. 17, 1930
He died as he lived-A
Christian
J. CUTHBERT HARE
SEPT. 28, 1879
APR. 18, 1936
ANNIE RAVENEL
MAR. 19, 1865
FEB. 28, 1954

References

Arrow, Kenneth J., Claire B. Panosian, and Hellen Gelband, eds. 2004. *Saving Lives, Buying Time: Economics of Malaria Drugs in an Age of Resistance.* Washington, DC: National Academies Press.

Ball, Anne Sinkler. 1984. Pinopolis: history of a pineland village.

Ball, Edward. 1998. *Slaves in the Family.* New York: Farrar, Straus and Giroux.

Ball, William. 1983. Letter to Anne Ball, May 22, 1983. In the possession of Sarah Cain Spruill.

Barker, Mrs. T. G. Letter to Charles Lucas. Lucas Papers, South Carolina Historical Society.

Berkeley County, South Carolina, Webpage. School pupils in Pinopolis School in 1924. www.rootsweb.com/~scberkel/.

Brewster, Lawrence Fay. 1947. *Summer Migrations and Resorts of South Carolina Low-country Planters.* Durham, NC: Duke University Press.

Brown, Mary (Dolly). History about places and people. In the possession of the author.

Bull, Elias Ball. Collected papers, South Carolina Historical Society.

———. New England City. Bull papers, South Carolina Historical Society.

———. 1962. Community and neighborhood names in Berkeley County, part II. *Names in South Carolina* 9 (Winter).

———. 1965. Community and neighborhood names in Berkeley County, III. *Names in South Carolina* 12 (Winter): 37–45.

———. 1979. Historic Preservation Inventory, Berkeley County.

Bull, Kinloch. Letter to Elliot West. Courtesy of Elliot West.

Cain Genealogy. Private papers of Sarah Cain Spruill.

Cain, Caroline. 1958. A sketch of the history of the plantations of Somerset and Somerton. Private papers of Sarah Cain Spruill.

Cain, Charles. 1973. Interview. Elias Ball Bull papers, South Carolina Historical Society.

Cain, William. Somerset Plantation, 1910–1940.

Carter, R., and K. Mendis. *Microbiology Reviews.* October 2002, p. 580.

Charleston (SC) News and Courier.

Childs, Arney Robinson. 1947. *The Private Journal of Henry William Ravenel, 1859–1887.* Columbia, SC: University of South Carolina Press.

Cote, Richard N. 1981. *Local and Family History in South Carolina.* Southern Historical Press.

Cross, J. Russell. 1985. *Historic Ramblin's Through Berkeley.* Columbia, SC: R. L. Bryan Co.

Davidson, Chalmers Gaston. 1971. *The Last Foray; the South Carolina Planters of 1860: A Sociological Study.* Columbia, SC: University of South Carolina Press.

Deas, Anne Simons. 1910. Two years of plantation life. Caroliniana Library, Columbia, SC.

Desowitz, Robert. 1991. *The Malaria Capers.* New York: W. W. Norton & Co.

Dickson, S. H. 1846. Essay on malaria: the proceedings of the agricultural convention on the State Agricultural Society of South Carolina from 1839–1845, inclusive. Columbia, SC.

Doar, David. 1908. A sketch of the agricultural society of St. James, Santee, South Carolina and an address on the traditions and reminiscences of the parish. Delivered before the society on July 4, 1907. Charleston, SC: Calder-Fladger Co.

Dubisch, J. 1985. Low country fevers: cultural adaptations to malaria in antebellum South Carolina. *Social Sciences & Medicine* 21 (6): 641–649.

DuBose, Samuel, and Frederick A. Porcher. T. Gaillard Thomas, ed. 1887. Reminiscences of St. Stephen's Parish. *A Contribution to the History of the Huguenots of South Carolina.* New York: Knickerbocker Press.

Dwight, H. R. 1921. *Some Historic Spots in Berkeley*. Charleston, SC: J. J. Furlong Printing House.

Edgar, Walter. 1984. *History of Santee Cooper, 1934–1984*. Columbia, SC: R. L. Bryan Co.

———. 1998. *South Carolina: A History*. Columbia, SC: University of South Carolina Press.

Fishburne, Anne S. 1969. *William Kershaw Fishburne, M.D.: Doctor to Hell Hole Swamp*. Columbia, SC: R. L. Bryan Co.

Gribbin, Emma Jenkins. 1968. Pinopolis from 1895 to 1911. Elias Ball Bull Collection, South Carolina Historical Society.

Grimball, Panchita Heyward. Pineland villages of Berkeley County. Elias Ball Bull Collection, South Carolina Historical Society.

Harper's Weekly 5, no. 213 (January 26, 1861): 54.

Haygood, Tamara Miner. 1987. *Henry William Ravenel, 1814–1887, South Carolina Scientist in the Civil War Era*. Tuscaloosa: University of Alabama Press.

Historic Resources of Berkeley County. 1990. Berkeley County Historical Society.

Irving, John B. 1969. *A Day on Cooper River*. Columbia, SC: R. L. Bryan Co.

Jervey, Susan R., and Charlotte S. Ravenel. 1921. *Two Diaries from Middle St. John's, Berkeley, South Carolina, February–May 1865. Journals Kept by Miss Susan R. Jervey and Miss Charlotte S. Ravenel at Northampton and Pooshee Plantations, and Reminiscences of Mrs. Mary Rhodes (Waring) Henagan*. St. John's Hunting Club.

Kirk, Francis Marion. Plantations of St. John's. Berkeley County Historical Society, www.rootsweb.com/~scbchs/plantation5.html.

———. Unpublished writings, courtesy of Videau Kirk Simons.

Leiding, Harriette Kershaw. 1921. *Historic Houses of South Carolina*. Philadelphia, London: J. B. Lippincott Co.

Louise, E. 2001. *Elizabeth Clevland Hardcastle, 1741–1808: A Lady of Color in the South Carolina Low Country*. Columbia, SC: Phoenix Publishers.

McCarrier, Herbert Geer, Jr. History of the missionary jurisdiction of the south of the Reformed Episcopal Church, 1874–1970. *The Historical Magazine of the Protestant Episcopal Church* 41, no. 2 and no. 3.

Merrens, H. Roy and George D. Terry. 1984. Dying in paradise: malaria, mortality and the percepted environment in colonial South Carolina. *Journal of Southern History*. L:4, November 1984.

Mills, Robert. 1826. *Statistics of South Carolina, Including a View of Its Natural, Civil, and Military History, General and Particular*. Charleston, SC: Hurlbut and Lloyd.

Minutes of Board of Trustees of Pinopolis Academy. Berkeley County Library, Moncks Corner, South Carolina.

Native Americans of North Carolina. *Microsoft Encarta Online Encyclopedia*.

Orvin, Maxwell Clayton. 1950. *Moncks Corner, Berkeley County, South Carolina*. [Charleston?].

———. 1961. *In South Carolina Waters, 1861–1865*. Charleston, SC: Nelson's Southern Printing & Publishing Co.

———. 1973. *Historic Berkeley County, South Carolina, 1671–1900*. Charleston, SC: Comprint.

Owens, M. A., PhD. 1983. Cousin Isaac Porcher, a vignette. *Transactions of the Huguenot Society of South Carolina*. No. 88.

Owens, Mildred. 1998. Just remembering. Privately published.

Pinopolis United Methodist Church History pamphlet, September 1986.

Porcher and Haskell Family Papers. Library of the University of North Carolina at Chapel Hill, Southern Historical Collection.

Porcher, Anne Allston. Papers, South Carolina Historical Society.

Porcher, Frederick A. (F. A.). 1887. *Historical and Social Sketch of Craven County, South Carolina, and History of the Huguenots of South Carolina*. New York: Knickerbocker Press.

———. 1921. *Upper Beat of St. John's Berkeley, A Memoir: A Contribution to the Huguenots of South Carolina.* New York: Knickerbocker Press.

———. 1943. Memoirs of Frederick Adolphus Porcher. *South Carolina Historical and Genealogical Magazine.* In 14 volumes: vol. 44, April 1943 to vol. 48, January 1947.

Porcher, John Henry. Personal diary, 1899–1910. In the possession of Martha Porcher Ravenel.

Porcher, P. C. Family correspondence. South Carolina Historical Society.

Porcher, Richard D. 1957. The history of Pinopolis. Unpublished.

Porcher, Walter Peyre, MD. Letter to Virginia Leigh Porcher. Anne Allston Porcher papers, South Carolina Historical Society.

Ramsey, David. 1809. *History of South Carolina from Its First Settlement in 1670 to the Year 1808.* 2 vols. Charleston.

Ravenel, Emily G. 1965. Letter to Florence Lucas. Private collection, Sarah Cain Spruill.

Ravenel, Henry W. 1947. *The Private Journal of Henry William Ravenel, 1859–1887.* Columbia: University of South Carolina Press.

Ravenel, Marion Rivers. 1991. *Marie Ravenel: From Childhood to China.* Charleston, SC: J. R. Rowell Printing Co.

Ravenel, Samuel Wilson. 1903. Christmas at Pooshee. Private letter, courtesy of Richard D. Porcher.

Ravenel, Thomas Porcher. Civil War diary. Charleston County Library.

———. Journal of Thomas Porcher Ravenel, 1846–1898. Unpublished, South Carolina Historical Society.

Savage, Henry. 1968. *River of the Carolinas: The Santee.* Chapel Hill: University of North Carolina Press.

Schlagenhauf, P. 2004. Malaria from prehistory to present. *Infectious Disease Clinics of North America* 18:189–192.

Silverman, Jason H., and Judith M. Andrews, eds. 2001. *South Carolina Then and Now.* Columbia, SC: South Carolina Department of Archives and History.

Smith, Henry A. M. French Jamestown. *South Carolina Historical and Genealogical Magazine* 9:220.

———. 1988. *Cities and Towns of Early South Carolina.* Reprint.

Smythe, A. T., H. R. Sass, Alfred Huger, Beatrice Ravenel, Thomas Waring, Archibald Rutledge, Josephine Pinckney, Caroline Pinckney Rutledge, Dubose Heyward, Katherine Hutson, and Robert W. Gordon. 1931. *The Carolina Low-Country.* New York: Macmillan Co.

Spruill, Sarah Cain. Private records.

Stevens, Margaret. Personal records, courtesy of Robert Stevens.

———. 1982. Pinopolis Village. *Names in South Carolina* 39 (Winter): 34–35.

Stoney, John Safford. Letters from Paris to his mother at Medway, 1856. Edited by John L. Tison Jr. University of Georgia, submitted to South Carolina Historical Society.

Taylor, Rosser H. 1942. *Ante-Bellum South Carolina: a Social and Cultural History.* Chapel Hill: University of North Carolina Press.

Thomas, Albert Sidney.1957. *Historical Account of the Protestant Episcopal Church in South Carolina.* Columbia, SC: R. L. Bryan Co.

Thomas, T. Gaillard. 1887. *A Contribution to the History of the Huguenots of South Carolina, Consisting of Pamphlets by Samuel Dubose and Frederick A. Porcher.* Republished for private circulation by T. Gaillard Thomas. New York: Knickerbocker Press. Reprinted by the Southern Printing Co., Mullins, South Carolina.

Towles, Louis P. 1996. *A World Turned Upside Down: The Palmers of South Santee, 1818–1881.* Columbia, SC: University of South Carolina Press.

Turnbull, Robert. 1956. *Bibliography of South Carolina, 1563–1950.* 6 vols. Charlottesville: University of Virginia Press, 1956–1960.

Tyler, David A. Memoirs. Elias B. Bull papers, South Carolina Historical Society.

Walker, Legare. 1910. *A Sketch of the Town of Summerville, South Carolina.* [Summerville, SC].

Waring, J. I. 1923. Waring family. *South Carolina Historical and Genealogical Society Magazine* 24.

Waterman, Thomas T. 1939. A survey of the early buildings in the region of the proposed Santee and Pinopolis reservoirs in South Carolina. United States Department of the Interior, National Park Service, Historic American Buildings Survey.

Wesley United Methodist Church, Pinopolis. Church archives.

West, H. Norman, Sr. History of First Baptist Church, Moncks Corner.

West, W. Edgar. 1953. Berkeley County Baptists. Edited by Wallace Edwards.

Wiley, William H. 1955. Tumuli in the Pinopolis Reservoir. Reproduced by the South Carolina Public Service Authority, 1961, 1968.

Wilson, Robert. 1907. An address delivered before the St. John's Hunting Club, July 4, 1907. Charleston, SC: Press of Walker, Evans, & Cogswell Co.

———. 1928. *Half Forgotten By-Ways of the Old South.* Columbia, SC: State Co. "In the Pineland" published in *Lippincott's Magazine*, October 1875.

Index

I

J

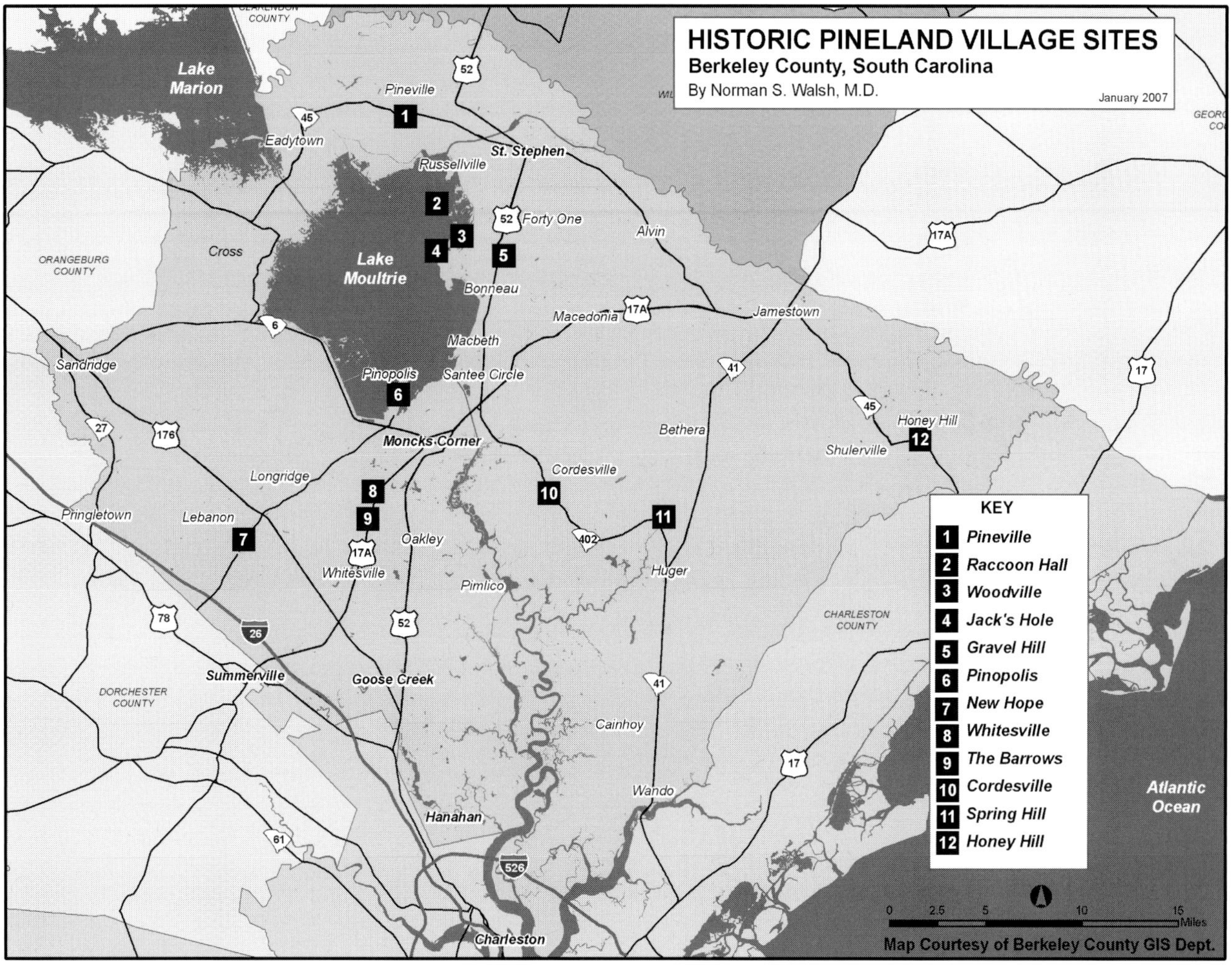

This map is adapted from a map created by Elias Bull titled "Pineland Villages." His map was published in *Historic Inventory of Berkeley County*, 1979.

Biographies

Norman Sinkler Walsh, MD, grew up in Pinopolis when it was, in many ways, still like a pineland village in the nineteenth century. He was educated in Berkeley County public schools and still considers himself a "country person." He practiced general surgery in the Charleston area from 1968 until 2001. His many interests include golf, hiking, long-distance cycling, woodworking, and photography. His interest in digital photography led to a desire to preserve visual images of the remaining historic homes in Pinopolis. He and his wife, Marcy Stoney Walsh, live in Summerville. They have four adult children and four grandchildren.

Cecy Guerry received a BA in English from Clemson University in 1972. She holds master's degrees in English and Secondary Education (English) from Clemson. Guerry taught high school English for seventeen years in South Carolina and Georgia. She also earned a Master in Library Media Services degree from Emory University in Atlanta, Georgia. She worked in public school libraries for sixteen years before retiring in 2005. She lives in Pinopolis, South Carolina, with her husband, George J. Davis III.

Additional Photographs and Information

Since the first printing of this book, significant information and photographs have been discovered. Of these, the most interesting and pertinent are included in this second printing.

Pinopolis' oldest natives were all over age eighty when interviewed for this book in 2005. In gratitude for their sharing fascinating stories and valuable information, we have added brief biographies.

Beatrice (Bea) Dennis Plyler was delivered by Dr. W. K. Fishburne in Pinopolis in 1912. She was a desk mate and shared an inkwell with Dr. Fishburne's daughter Emily (Mrs. Ben Scott Whaley), who later became a well-known Charleston author and gardener. Mary Causey, daughter of Sheriff Causey, was Bea's first grade teacher. Miss Causey married, and Mrs. Edwards, mother of Margaret Edwards Stevens, replaced her. Bea's father, Senator Ed Dennis, was shot and killed in Moncks Corner in 1930. Seventy-five years later, she still wonders "why" and has strong emotional feelings regarding his murder. Bea still has vivid memories of life in Pinopolis as a child. After college, she first taught school in Cross, where Francis Marion Kirk was the principal. She later married John Plyler, who became president of Furman University. On the morning that we visited her at home in Greenville in August 2005, she looked spry and was still driving her automobile. At age ninety-three, she took us to lunch in town, read the menu without glasses, and had enough energy to talk until 3:00 p.m.

Harold Huxford Harvey Jr. was born in 1920 in Pinopolis. He attended Pinopolis School until the third grade or about 1929. He remembers that when he went to Moncks Corner for the fourth grade his teacher was Mrs. Fultz. In 1946, sixteen years later, Mrs. Fultz was my fourth grade teacher. Some of Harold's Pinopolis friends were Markley Dennis, Davis Hare, and Carleton Massalon. Harold worked for Santee Cooper from its beginning until 1951, when he and his wife Frances ran Harvey's five & dime on Main Street. Harold and Frances sold their business in 1979. Harold served in the U.S. Army from 1942 until 1946. He and Frances married in 1946. For those who remember the Lollipop Trolley on Main Street, Frances served ice cream there from 1946 until 1950. I remember that one could get a cone for five cents per scoop. Harold enjoyed keeping cows and doing light farming as a hobby.

Mildred (Mickey) Carson Owens was born in Pinopolis in 1917. Her parents were Frank Carson and Carrie McClure Abby. Her grandfather was Jacob Carson; he donated the old organ that is still used in St. John's Baptist Church. Mickey remembers first grade at the Pinopolis School where she walked to and from daily, bringing lunch. Mickey and Mary Brown were lifelong friends. For a time, Mickey was bookkeeper for Gippy Plantation when Mr. Carpenter managed it. She then worked for Berkeley County. In 1940, she married David Owens. From 1990 through 1993, Mickey wrote a series of articles about living in Berkeley County, which were published in the *Berkeley Independent*. In 1998, the stories were compiled and published in her book *Just Remembering, Old Pinopolis, Moncks Corner, and Berkeley County.*

Mary Rebecca Sheppard Brown (Dolly) was born in Pinopolis in 1917. Her story is told beginning on page 137 of this book, and she lived to see her story published in this book. In February 2007, the staff of the *Post and Courier* interviewed and photographed her. She died on February 19, 2007, just five days before the *Post and Courier* article was published.

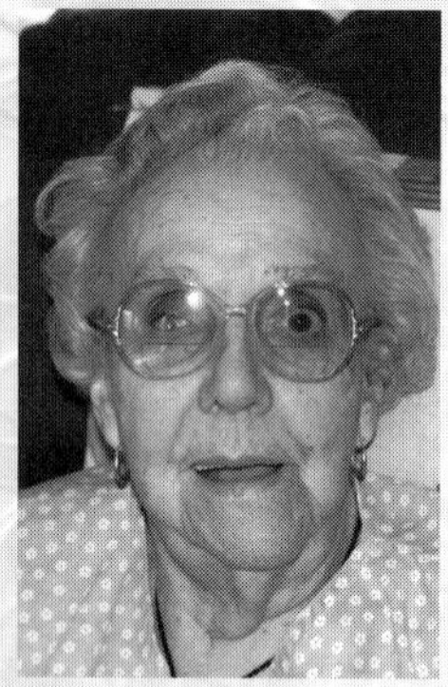

Julia Parker Young was born in Oakley in 1916 and moved with her family to Pinopolis in 1918 after her father, Samuel Pledger Parker, died. She first lived in the house built by L. H. Harvey (not listed in the first edition) near the Episcopal Church. From there, they moved to the E. O. Hall house, which had been in her mother's family (Guyton). Around 1930, the family moved to the John T. Grant house. The Parker family owned this house until September 2000. Julia went to the Pinopolis School until it closed. She became an accomplished pianist and played for many dances in the village. She attended the public school in Moncks Corner until the ninth grade, when her mother sent her to Memminger School in Charleston where she graduated in 1930. She boarded in Charleston along with her first cousins, Grace Smith (Mrs. Markley Dennis) and her sister Ruth Smith (Mrs. David Page). Julia became an excellent tennis player under the direction of Emily Fishburne (Whaley). She excelled at tennis during her years at Lander College.

After Julia married Thomas Young, her mother, Louise Guyton Parker, gave the Robert Lindsay house to her in 1947. She lived in this house for the next sixty years. Julia taught home economics at Macedonia and Berkeley elementary schools. She worked for twenty years with the Berkeley County Welfare Department. She founded Punch and Judy Kindergarten and operated this from her home for thirteen years. She was a member of the Pinopolis Book Club, Pinopolis Garden Club, and the Pinopolis United Methodist Church. She died on June 30, 2007.

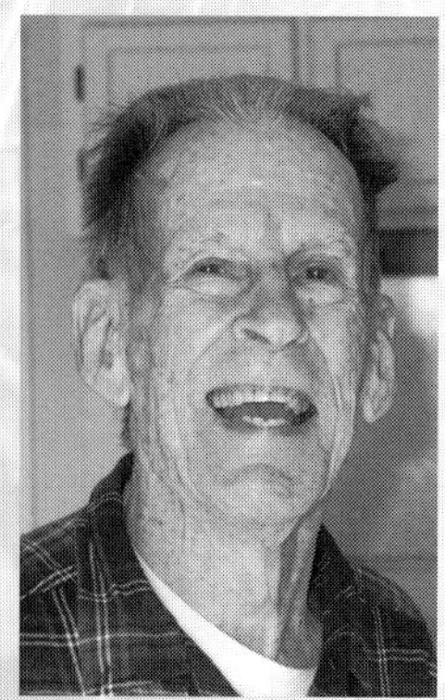

Henry Ravenel Dwight Jr. was born in Pinopolis in 1911 in the second home built by Frederick A. Porcher. He attended Pinopolis School and was proud of a Latin award that he won in 1925. Henry noted that his father designed the Pinopolis Public School. After graduating from the Citadel in 1934, Henry went to Sandridge as the school principal. After that, he worked with the Civilian Conservation Corps at Witherbee where my father also worked as a physician. In 1941, he joined the U.S. Army and participated in the invasion of North Africa. When the war ended in 1945, he went to Charleston and had a long and successful career in real estate. He was active in the French Huguenot Church and served as president of the South Carolina Historical Society.

L. H. Harvey/Connie Murray House. We studied this image from the beginning while researching this book. Although found among other Pinopolis photographs, no one could identify the subject of this photo as a Pinopolis house. Written at the bottom was "from the cupola." The only cupola in Pinopolis was at the top of the public school building. We gave up the search. However, in March of 2007 after the book was printed, we approached Harold Harvey, age eighty-seven, for a second time with this image. He said: "Well, let me just walk up the stairs of that old public school building [as it was around 1930] and see what I can see. I'm in the cupola now, and I see that house. I remember that it was built by Lewis H. Harvey for his daughter Connie Harvey [Murray]." This now all comes together. In our first interview with Julia Parker Young in September 2005, she described living in a house in this area, but nothing that we knew fit her description. Now we know that this is the house where Julia came with her mother and siblings after her father, the train depot manager at Oakley, died in 1918 of cholecystitis. Julia remembers that the house was "near the Trinity Episcopal Church." The house site was just north of the Pinopolis School and west of the Kate Harvey house. Harold Harvey also remembered that the house burned. We examined this site but could not find any foundation remains. Connie Harvey Murray later built a brick house slightly to the north and west of this site.

St. John's Hunting Club meeting, May 1908, at Indianfield Plantation. Left to right beginning at top: Allen Jervey, Turner Logan, John St. Claire White, Faber Porcher, Edwin Dubose, Henry Ficken, Percival R. Porcher II, Henry Lucas, James R. Cain, C. Stevens White, John St. Clair White Jr., Joseph E. Jenkins, and Dan Jenkins; second row: Henry Ravenel Dwight and Walter Peyre Porcher; third row: Col. James Armstrong, Richard Yeadon Dwight, J. Calhoun Cain, and Isaac de Cherigny Porcher; fourth row: Thomas Porcher Ravenel, John Henry Porcher, and William Walker. *Courtesy of Martha Mitchell Porcher Ravenel.*

St. John's Hunting Club, May 1974. Members from left to right: Francis Cain, Henry Ravenel, William E. Martin, Joseph P. Cain, Charles Cain, Dr. Rene Ravenel, Dr. Jervey Ravenel, Henry Ravenel Jr., Dr. J. N. Walsh, Dr. Norman S. Walsh, Dr. James Ravenel, W. Edgar Davis, Lucas Gaillard, John Porcher Bryant, Dr. Sam Ravenel, William Cain, Henry Porcher, Henry R. Dwight, unidentified, unidentified, and James Jervey Ravenel Jr.

Sarazins Plantation House. William Henry Johnson (1871–1934) made this sketch of the Sarazins plantation house. It is in volume one of his scrapbook, which covers the years 1920–1933. A description of the plantation is on page 72. *Courtesy of the South Carolina Historical Society.*

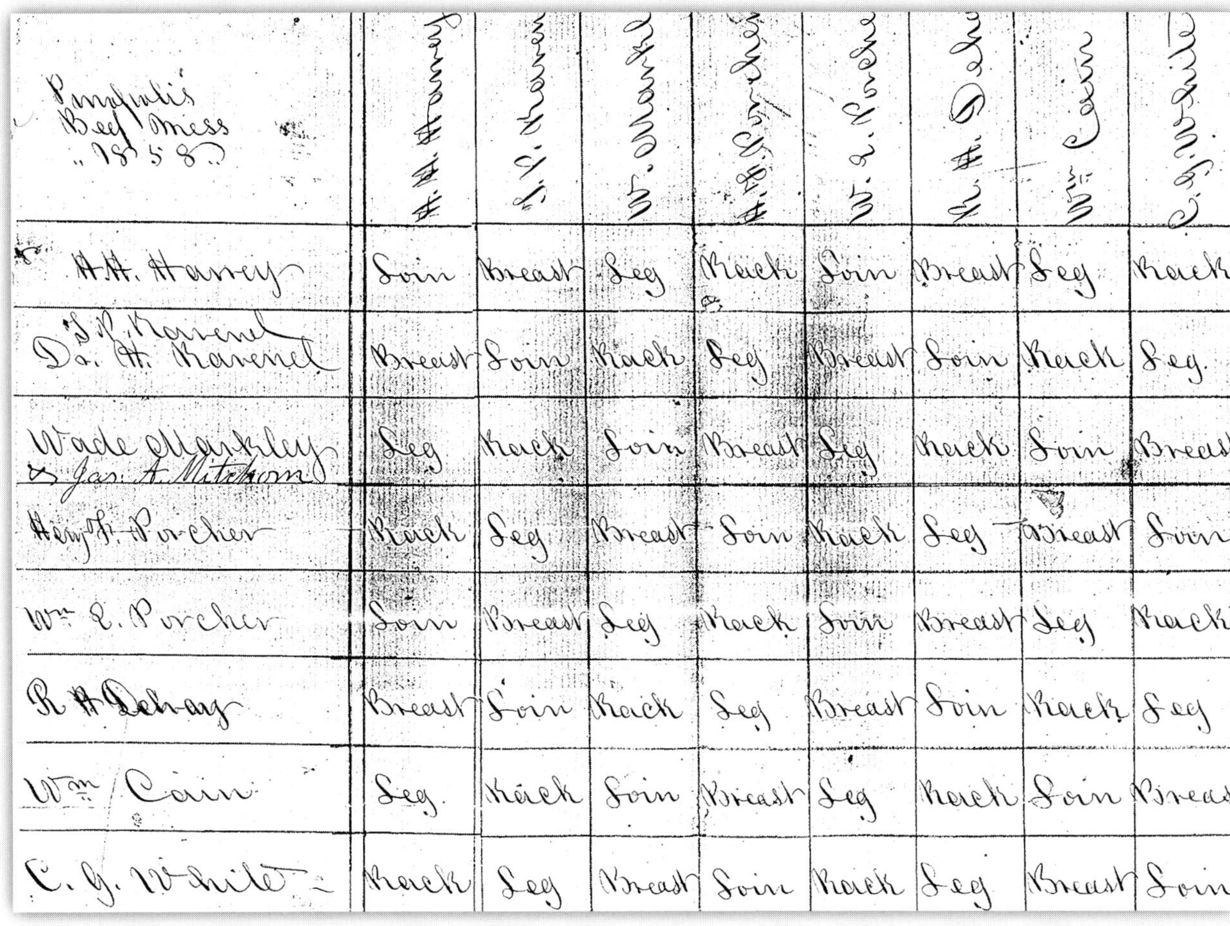

Pinopolis Beef Mess 1858	H. H. Harvey	T. P. Ravenel	Wm. Markley	H. F. Porcher	Wm. P. Porcher	R. H. Dehay	Wm. Cain	C. G. White
H. H. Harvey	Loin	Breast	Leg	Rack	Loin	Breast	Leg	Rack
T. P. Ravenel / Dr. H. Ravenel	Breast	Loin	Rack	Leg	Breast	Loin	Rack	Leg.
Wade Markley & Jas. A. Mitchom	Leg	Rack	Loin	Breast	Leg	Rack	Loin	Breast
Henry F. Porcher	Rack	Leg	Breast	Loin	Rack	Leg	Breast	Loin
Wm. P. Porcher	Loin	Breast	Leg	Rack	Loin	Breast	Leg	Rack
R. H. Dehay	Breast	Loin	Rack	Leg	Breast	Loin	Rack	Leg
Wm. Cain	Leg.	Rack	Loin	Breast	Leg	Rack	Loin	Breast
C. G. White	Rack	Leg	Breast	Loin	Rack	Leg	Breast	Loin

A Pinopolis beef mess table from 1858. Because there was no refrigeration, residents could only keep enough meat to feed their families for a day or two. They formed compatible cooperatives to kill one of their cows and divided the meat so that each person got a different cut each time. Each family killed a cow as their name came up in the rotation. Names in this table are H. H. Harvey; T. P. Ravenel and Dr. H. Ravenel; Wade Markley and Jas. A. Mitchom; Henry F. Porcher; Wm. P. Porcher; R. H. Dehay; Wm. Cain; and C. G. White. Courtesy of the South Carolina Historical Society.

William Cain (1792–1878), one of the participants in the 1858 Beef Mess, was very prominent in the beginning of Pinopolis and probably deserves an equal place with F. A. Porcher and Dr. Morton A. Waring in the history of the village. Cain is mentioned on pages 89–92. In May 1837, Thomas Porcher of White Hall plantation sent a long letter to his brother Peter Cordes Porcher, who was studying medicine in Paris. In that letter, Porcher mentions that he has moved a home from Goshen plantation to "the Pineland of Doctor Waring and Mr. Cain." This indicates that Cain was active in Pinopolis as early as 1837 and that the Somerset summer house may have been built in 1837 or earlier.

William Cain attended the Pineville Academy and graduated from the South Carolina College, now the University of South Carolina, in 1812. He taught at the Pineville Academy for two years. He was a vestryman at Black Oak Episcopal Church. In public service, he was a justice of the peace; justice of the quorum; magistrate; state representative (1820–1821 and 1828–1833); state senator (1841–1846); presidential elector (1844); lt. governor (1846–1848); member of Secession Convention (1860–1862); signer of the Ordinance of Secession; Commissioner of Free Schools; Commissioner of Public Buildings; and a trustee of South Carolina College. His mail was addressed to the Honorable William Cain. William Cain died in Pinopolis.

Pinopolis Residents 1853. In 1853, Thomas Porcher Ravenel (age 29) wrote a letter to his brother, Rene (age 27), who was studying medicine in Paris. This Rene was Dr. Rene Ravenel of the Pooshee summer house (page 64). In this letter, Thomas listed all of the people who were living in Pinopolis at the time. Their ages are in parentheses.

In the Chapel Hill summer house were Thomas Porcher Ravenel (29), Elizabeth (Liz) Margaret Wilson Ravenel, Henry LeNoble Ravenel (9), Samuel Wilson Ravenel (8), Thomas Porcher Ravenel (6), and Bessie Ravenel (1).

In the William Henry Ravenel (the botanist) house: Mrs. Solomon Clarke (53), Anne Camilla Clarke (18), Susie Elizabeth Clarke (16), Mr. and Mrs. Iley Coleman (Mary Jane Clarke, 24 and Iley Coleman, 37).

In the Sarazins summer house: Aunt Isabella Peyre Porcher (50), Louisa Porcher (23), Alexander (22), Julian (25), Bill (30).

At Bunker Hill: Mrs. Press McPherson Smith (66) and niece Miss Susan Dubose (47).

In the Whitehall summer house: Aunt Catherine Gaillard Porcher (58), Charles Lucas (24) & Liz Porcher (28), John White (33), and Tom Porcher (26).

In the Somerset summer house: Mr. Cain (61) and William Cain (19).

In the Northampton summer house: Henry LeNoble Stevens (26) & Henrietta Stevens Gaillard Stevens (24), and Percival Porcher (24).

In the Wampee summer house: Capt. Robertson (57) & Mary Macbeth Robertson (53).

In the Chelsea summer house: Dr. Waring (44), John Waring (20), Edith Waring (15), Rebecca Waring (18), Morton Waring (21), Old Mr. Waring (69), and Archibald Waring (19) are visiting.

In the Ophir summer house: Uncle Henry Porcher (34), Grandmother Elizabeth Sinkler DuBose Porcher (65), Isaac DuBose Porcher (21), Jane Porcher DuBose (45) boarding at Ben Ravenel's.

In the Woodboo summer house: John Mazyck.

In the Kate Harvey (Hogswamp) summer house: William Dennis (51).

To purchase additional copies of this book, please contact:

The Collectors' Corner
308 East Main St.
Moncks Corner, SC
29461-0968
ph: 843 899-1886

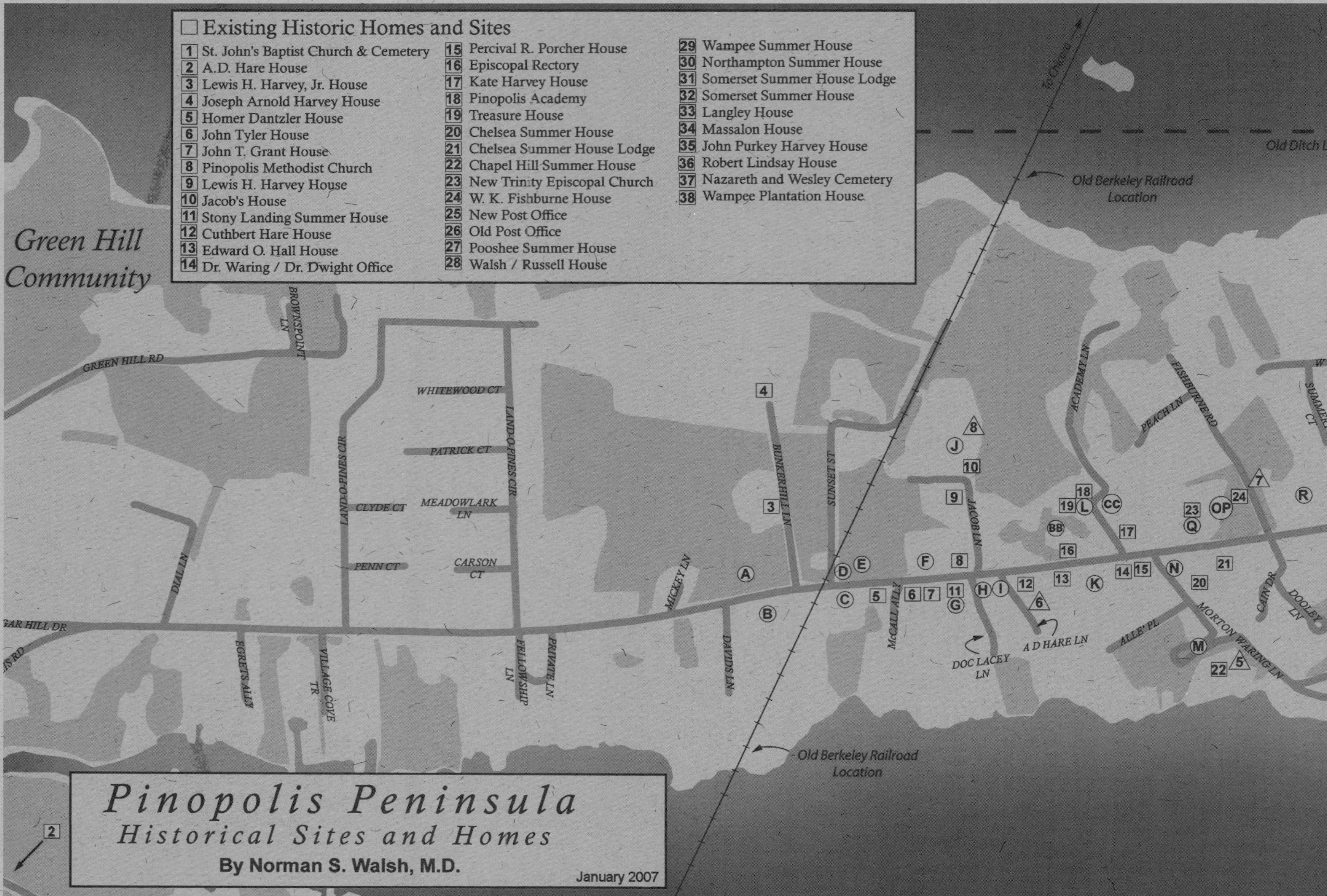
Existing Historic Homes and Sites
1 St. John's Baptist Church & Cemetery
2 A.D. Hare House
3 Lewis H. Harvey, Jr. House
4 Joseph Arnold Harvey House
5 Homer Dantzler House
6 John Tyler House
7 John T. Grant House
8 Pinopolis Methodist Church
9 Lewis H. Harvey House
10 Jacob's House
11 Stony Landing Summer House
12 Cuthbert Hare House
13 Edward O. Hall House
14 Dr. Waring / Dr. Dwight Office
15 Percival R. Porcher House
16 Episcopal Rectory
17 Kate Harvey House
18 Pinopolis Academy
19 Treasure House
20 Chelsea Summer House
21 Chelsea Summer House Lodge
22 Chapel Hill Summer House
23 New Trinity Episcopal Church
24 W. K. Fishburne House
25 New Post Office
26 Old Post Office
27 Pooshee Summer House
28 Walsh / Russell House
29 Wampee Summer House
30 Northampton Summer House
31 Somerset Summer House Lodge
32 Somerset Summer House
33 Langley House
34 Massalon House
35 John Purkey Harvey House
36 Robert Lindsay House
37 Nazareth and Wesley Cemetery
38 Wampee Plantation House
Green Hill Community
To Chicora
Old Berkeley Railroad Location
Old Berkeley Railroad Location
GREEN HILL RD
BROWNSPOINT LN
WHITEWOOD CT
PATRICK CT
LAND-O-PINES CIR
LAND-O-PINES CIR
CLYDE CT
MEADOWLARK LN
PENN CT
CARSON CT
DIAL LN
EGRETS ALLY
VILLAGE COVE TR
FELLOWSHIP LN
PRIVATE LN
DAVIDS LN
MICKEY LN
BUNKERHILL LN
SUNSET ST
McCALL ALLY
JACOB LN
DOC LACEY LN
A D HARE LN
ACADEMY LN
PEACH LN
FISHBURNE RD
ALLE' PL
MORTON WARING LN
CAIN DR
DOOLEY LN
SUMMERTON CT
Pinopolis Peninsula
Historical Sites and Homes
By Norman S. Walsh, M.D.
January 2007